ENTRY PARADISE

GERHARD SELTMANN – WERNER LIPPERT
(EDS.)

ENTRY PARADISE

--- NEW WORLDS OF DESIGN ---

BIRKHÄUSER – PUBLISHERS FOR ARCHITECTURE
BASEL · BOSTON · BERLIN

BEAUTY AND ECONOMICS

SPACE AND COMMUNICATION

OBJECT AND NATURE

ENTRY PARADISE

NEW WORLDS OF DESIGN

THE ZOLLVEREIN COAL MINE – A PLACE OF RENEWAL

GERHARD SELTMANN

------- A PAUSE FOR THOUGHT BEFORE ENTERING PARADISE? -------

The idea for the *ENTRY Paradise* book was inspired by "ENTRY2006 – How will we live tomorrow?", an exhibition on the future of design at the world heritage site of the Zollverein coal mine in Essen. Yet another exhibition that exploits the charm of old industrial buildings to enhance its own appeal?

Far from it: the shaft complex Zollverein XII, built in 1932 by architects Schupp and Kremmer, is not just the only industrial complex in the world constructed entirely in Bauhaus style. Until its closure in 1986 Zollverein XII, with a daily output of 22,000 tons, was also a gigantic machine, the most efficient coal mine in the world.

Architects and engineers collaborated closely on the construction of Zollverein XII to produce an architecture that deliberately opposed the contemporary aesthetics of industrial construction and was geared entirely towards functionality and maximum performance.

Yet for decades on end, it was the plant and architecture that shaped open-cast coal mining facilities. The world heritage site Zollverein is then a symbol for the magical effect of organised interaction between designers and technicians. That is the reason why architects like Lord Norman Foster, OMA or the Sanaa group have contributed to the reconstruction and structural development of the Zollverein complex. It is also why the Zollverein locations are home not only to the red dot design museum but approximately 60 creative service companies as well. And it is for this same reason that, in 2006, the Zollverein is the "born" venue for an exhibition on perspectives and visions in design and architecture.

ENTRY2006 focuses on three questions: how do the connections between research results, design experiments and the day-after-tomorrow's products arise? How does mutual inspiration between the creative disciplines work, from landscape design and architecture all the way to product and communications design? And where is design heading when research takes possession of what has hitherto been invisible – in biotechnology, nanotechnology or genetic research?

An exhibition can make important contributions to resolving such questions. It displays thrilling objects, presents designers and their biographies, points out lines of development, visualises connections. But an exhibition also lives from its visual presence – its impact fades as the event recedes into the past. The question as to the future of design, however, continues to be of huge social relevance. It is an issue not only for designers and companies, but the public agenda as well.

In the future too, of course, new products or architectures will continue to make everyday life easier, provide greater comfort, create new ways of living. The designing disciplines remain the breeding ground for innovation and the basis for commercial success.

Yet that is precisely why it is so important to consider the social and ethical dimensions of a development that uses design intervention to substitute artificial nature for the real thing, makes humans and the most intimate areas of their lives designable at nano-level or is giving the next generation of robots an increasingly human appearance. That is why, even during the conception phase, the curators of the ENTRY exhibition – Francesca Ferguson, Werner Lippert, Ellen Lupton, Peter Reed, Alexander von Vegesack and Peter Wippermann – agreed with the Zollverein exhibition company that an "accompanying book" on the topic was indispensable.

My very special thanks for their work on this publication are due to Werner Lippert, Thomas and Heike Edelmann as well as Katja Aßmann. *ENTRY Paradise* points out new and forgotten connections between science and design and sheds light on the relationships between the feasible and the desirable. It is not about naïve futurism or a fashionable fear of the future, but about an opening balance for a new age of design.

The authors of the book warn us not to leap to hasty conclusions about traditional design certainties. Norbert Bolz presents future possibilities for intervening in almost every aspect of life and focuses not only on the enthusiasm but the collective fear that goes hand in hand with new, nano-scale design options. He envisages a new job description for designers, as "shepherds of being" or "wise stewards of the planet".

This book brings together some important texts about the "redesigning" of our world. Some of them are published here for the first time. Other contributions are already familiar from a different context but have been included because of their significance. Sceptics like American computer expert Bill Joy also get to have their say. He calls for the radical renunciation of nanotechnologies and the self-restriction of knowledge. Ulf Poschardt's essay investigates our ambivalent relationship to mutants, replicants and androids.

Jan van Rossem reminds us of the complexity of design and the balancing act between commercial success and the often surprising suggestions designers come up with. Wolfgang Pauser explains why, if it wants to be convincing in the long term, emotional design must refrain from launching a crass attack on our cerebellum.

Ellen Lupton investigates the growing significance of natural and artificial skins for things and presents examples from architecture, product design and the new media. Peter Reed gathered statements from leading landscape designers and used them to formulate a changed notion of landscape and space. Koert van Mensvoort explains why, since we have long been transforming nature with our own design ideas, our traditional concept of nature requires revision.

Stem cell researcher Frank Edenhofer reminds us of the discovery of the cell as the raw material for all life. Intervention by designers can be seen either as a miracle cure or a threat.

In Europe, the debate about future dimensions of design largely takes place in specialist circles. It would seem as if the USA and Canada have been quicker to grasp the social relevance of the issue. This may partly be due to the fact that the associated fields of research are clustered in North America.

ENTRY Paradise is an attempt to gather carefully selected international standpoints on a single platform so as to provide impetus for the continuing debate. Perhaps it makes sense to pause for thought before entering the paradise of tomorrow's design possibilities and discuss the "instructions" for using them.

WERNER LIPPERT – PETER WIPPERMANN

-------ENTRYPARADISE-------

ENTRYPARADISE is an exhibition project about the world of design. It will explain how design has developed, where it stands today and what its task might be tomorrow. Ever since the Industrial Revolution, the work of product designers had focused on the question "how do we want to live tomorrow?" – their vision of the future. In view of today's revolutionary technologies, this is a question we have to ask ourselves again. Information, nano and biotechnology are changing our everyday lives. For besides the products we surround ourselves with, our natural surroundings are becoming deliberately designable as well.

Design is about to undergo a paradigm shift – the extent to which new technologies have been intervening in the constructive, material, aesthetic and social practice of both architecture and design since the nineties is unprecedented. Today design starts at the level of the atom. We are drifting into the world of the invisible: virtual realities, nano- and biotechnology are increasingly influencing our aesthetics and providing new construction kits for our reality. Information and communication design are ensuring the controllability of highly complex connections and giving rise to virtual social systems. Design is becoming immanent to being, and the experience of design will be both physical and metaphysical. Design is becoming invisible, and design is making things visible. Design promotes a better world and is based on the dream people have of themselves.

DESIGN AS A COMPETITIVE FACTOR

It is surprising that the German business world still underestimates the value of design. According to a 2002 survey by Ernst & Young, 93 per cent of all German companies

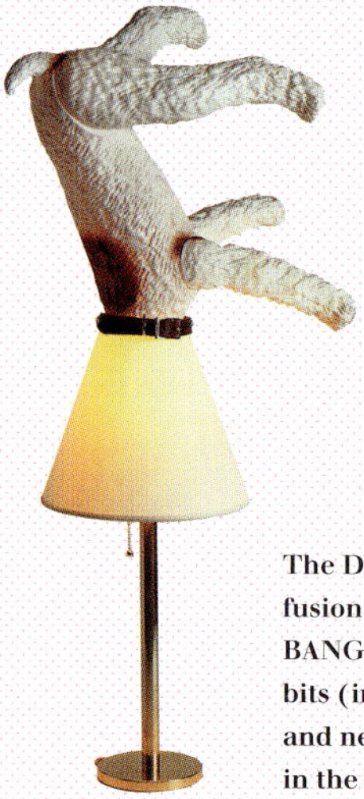

The DogLamp (karlssonwilker, NYC) visualised the fusion of two originally different objects. Under the motto BANG Design, ENTRYPARADISE offers an idea of how bits (information technology), atoms (nanotechnology), and neuro and gene technology can change our world in the future.

trust in the innovativeness of their products but only 13 per cent see design as a product benefit for the consumer. The majority of German companies largely abstain from using design to increase their value.

The success of the Apple iPod is a good illustration of Germany's strengths and weaknesses as an economic and design location. The basis of its technology, the MP3 compression standard, was developed at the Fraunhofer Institute. The Ulm School of Design was the inspiration for the iPod's design. Yet it was American entrepreneur Steve Jobs and his company, Apple, that turned the assertive linking of technology and aesthetics into a competitive advantage. The creativity of the design consisted in presenting online databases and portable hard drives as a cultural game. Its success cannot be explained solely on the grounds of its technological innovation. The iPod became an international fashion statement, the fetish of an urban lifestyle. A product development geared towards consumer needs and the skills required to market it at global level are America's strengths. Germany's economic weakness is that it fails to market its own innovative technologies and cultural traditions together.

The product as starting point was yesterday. Tomorrow the consumer will be the focal point for enhancing value. The emerging knowledge economy relies on digital self-determination in real time. The individual linking of what consumers want with the solutions offered by the databases, from music downloads all the way to personal medication, will shape the economy of the 21st century.

In future, both designers and their clients will have to dissociate themselves from the task of styling isolated products and adopt a holistic approach to their work. It is worth bearing the origins of industrial design in mind. The Bauhaus movement wanted to use the structural change from agricultural economy to industrial society to design a better quality of life. Even then, designers wanted to use new industrial manufacturing tech-

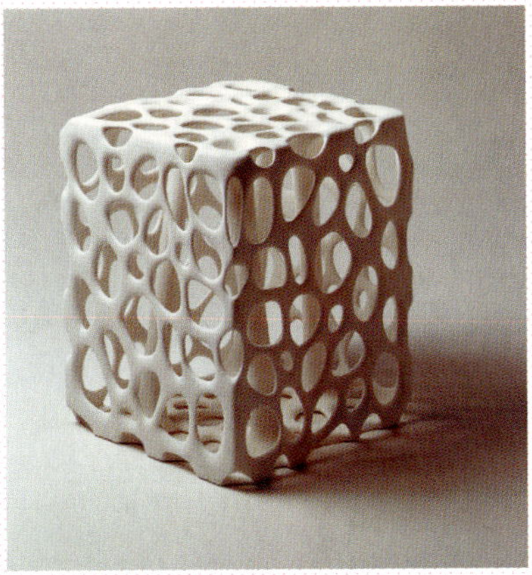

"Solid S1," the stool by French designer Patrick Jouin, is furniture, object and architecture all at the same time. As a 3D printed model, it symbolises the digital interface between design and production.

nologies to create better living space. The new technological situation makes it essential to rediscover this holistic interpretation of design.

BUSINESS DESIGN

Now that industrial production processes have been re-engineered or outsourced to China and India and controlling has improved the latest quarterly results, engineers, managers, bookkeepers and lawyers are running out of ideas. Now it's designers' turn to help. The famous German founder Hasso Plattner, chairman of the supervisory board of German software giant SAP, was the first to recognise the opportunities and acted on his insight in October 2005. He donated 35 million dollars to Stanford University in Silicon Valley, which subsequently opened the Hasso Plattner Institute of Design, or "d.school", on its campus. It's motto: "We believe having designers in the mix is key to success in multidisciplinary collaboration and critical to uncovering unexplored areas of innovation."

Designers see, understand and network the world illogically, intuitively and creatively. They come up with solutions conceived with man as their starting point. Their imagination is not moulded by the technological and economic limitations. They are capable of giving concrete shape and form to their daydreams. Using their emotional skills, designers – in collaboration with scientists from computing, medicine and physics – will develop new, interdisciplinary fields of application for materials and processes, with the result that basic research can be used for future innovations far more quickly. These new alliances between design strategies and business will lead to enduring changes in the way the knowledge economy creates value.

In Great Britain, the London Business School has joined forces with the University of the Arts London to found the Centre for Creative Business. New Zealand, Finland and Korea have launched government programmes for promoting design so as to hold their

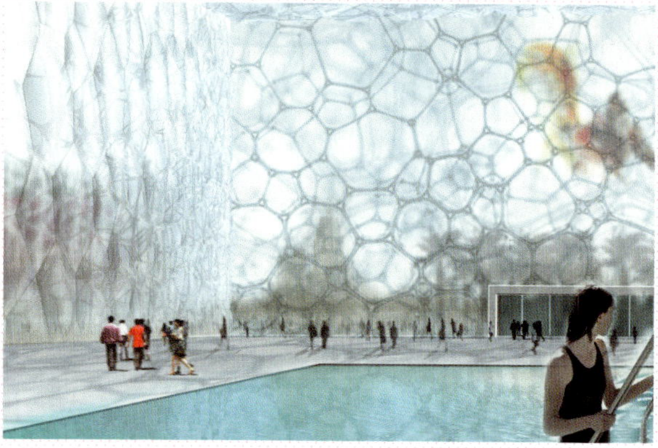

"Watercube", the design for an indoor swimming pool (PTW Architects, CCDI, Arup) for the 2008 Olympic Games in Beijing (China), is based on the principle of a soap bubble in its design and construction. Emotional desire shapes our reality. Architecture practices image building in two ways. Pictures we can enter into give our dreams a realistic dimension.

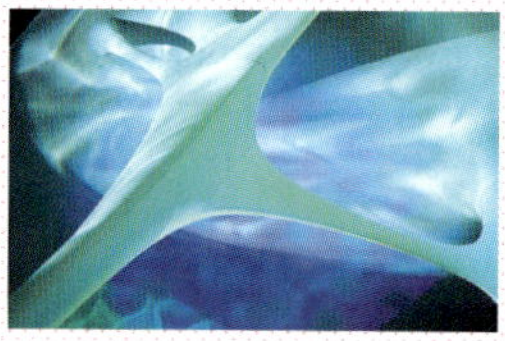

The design of the ENTRY pavilion "Visual Power" also originates from PTW Architects, Australia. Its organic structure, like that of a cell, follows natural courses of energy.

own in the face of global competition. The New Zealand Institute of Economic Research established that the 25 biggest economic powers are also the 25 best design nations. In Germany too, commerce and design will seek new ways of collaborating. The ENTRY-PARADISE exhibition is an initial attempt to raise awareness of the opportunities and risks of this new approach to value creation.

CHANGING VALUES

ENTRYPARADISE will be presented on four storeys of the converted boiler house of the Zeche Zollverein coal mine. The boiler house is a monument to the way values are changing as we move from an industrial to a knowledge society. Architect Rem Koolhaas' redesign of the boiler house translates the radical changes in production and architecture into a sensory experience. The Zeche Zollverein, designed in 1927 by Fritz Schupp and Martin Kremmer, was the most modern industrial mining complex in the Ruhr. In close collaboration with the mining engineers, Schupp and Kremmer developed their idea of interaction between machine and architecture, translating the fundamental principle of modernism into practice by insisting that the function determined the form. The Zeche Zollverein was built in the tradition of New Realism. The architects drew up their plans from the perspective of optimised working life. This new quality in the organisation of the working world became a symbol in the architecture. The Zeche Zollverein's rededication as Essen's design location is a sign of the structural change currently in progress. Now a World Heritage Site, the Zeche Zollverein stands for a new way of thinking in architecture and design, for conversion, renaturisation and new forms of working and living.

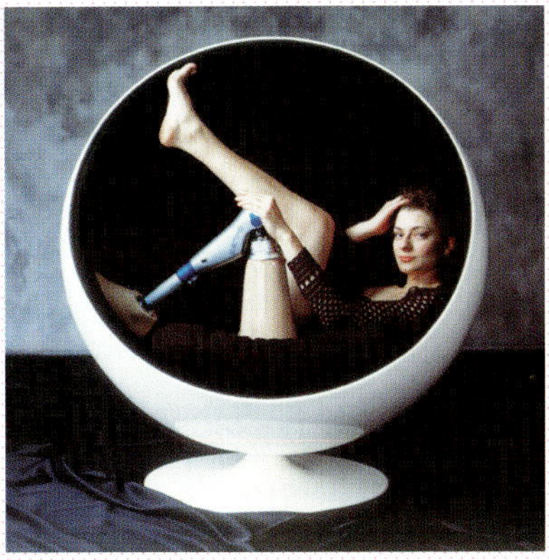

The C-Leg (manufacturer: Otto Bock) is the world's first knee-shin system to be controlled entirely by microprocessor. The C-Leg uses sensors and a microprocessor to automatically adjust to individual walking situations in real time and is the best approximation of natural walking to date.

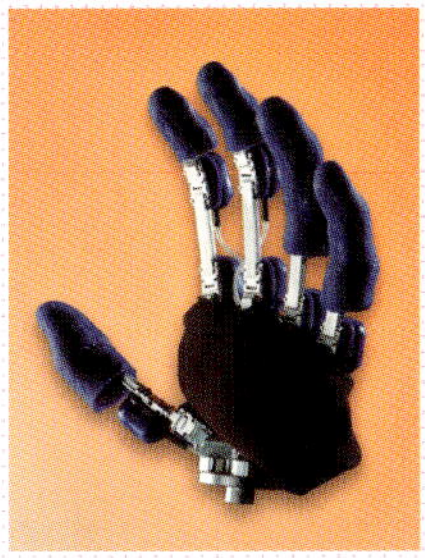

The demonstration model of a "fluid hand" was developed at the Institute for Applied Computer Science in Karlsruhe. Miniature electro-mechanical valves, sensors and energy storage make this hand prosthesis lighter. Many elements of its construction are integrated in the fingers and contribute to their movability.

BANG DESIGN

Our environment is becoming totally designable: nanotechnology is shaping smells and tastes, biodesign is creating surfaces and prostheses, information design is solving complicated procedures in black boxes and creating social networks. But it is the fusion of various new technologies that can look forward to the brightest future. We might call this new mega-science BANG: it operates at the level of bits (computing), atoms (nanotechnology), neurons (cognitive neurosciences) and genes (gene technology) and in related areas functioning at atomic and molecular level. The integration of nano and biotechnology is referred to as "nanobiotechnology". It permits the fusion of living and non-living material and the construction of hybrid organisms and products.

VISUAL POWER

Design in context with media – "creativity as symbol"

Today new information processing technologies are determining the development of design and computers are expanding its possibilities and forms. CAD (Computer Aided Design) gives designers unlimited creative freedom. Almost anything that can be drawn on a computer can be manufactured or built as well. The data set of the design has become the basis for production. Virtual crash tests have already proved their worth in aircraft and automobile construction. The viability of the design can be tested more effectively in the second reality of simulation than in the first reality of real things.

The paradox of this technological development is that the physical presence of the computer, the hardware, has provided no impulses for design at all. Thus the black box containing the software is supposed to be forgotten, which is why retro trends are determining today's design utopias. For designers and consumers alike, paradise lies in the past. Time islands are being settled. Whether it's the fifties, seventies or eighties doesn't

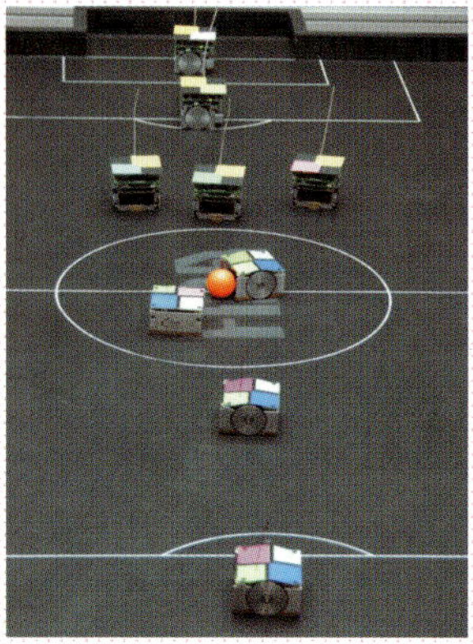

Teams of small, highly flexible cubes with two wheels, electromotors and radio receivers play against each other in a robot football match (here at the Technical University in Vienna). Colour codes identify the players, who play with an orange golf ball. A camera, installed above the playing field, transmits the current details of the game to a host computer that the players control via radio signals.

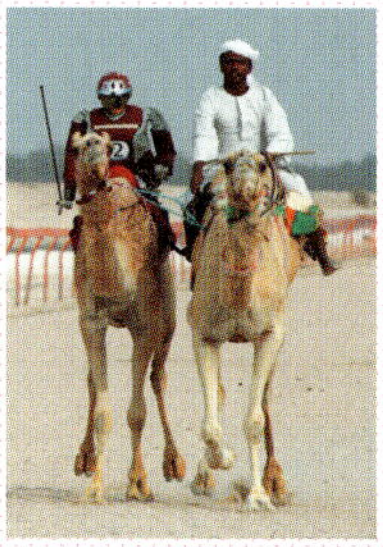

Robots have recently replaced small children as jockeys at traditional camel races in the Qatar Emirate. The electronic riders weigh about 30 kilogrammes, have a body-like shape, and can be controlled via voice or joystick.

really matter, for every style epoch has its own appeal. Deconstructivism as in the works of architect Zaha Hadid or Coop Himmelb(l)au, retrofuturism as in the designs of Jérôme Olivet or the visions of bionic architecture by Toyo Ito and Andrea Branzi are all enjoying major success.

But the ideas of the Bauhaus movement and the Ulm School of Design are experiencing a renaissance as well. Under computer conditions, however, the "form follows function" dogma is mutating into pure ornament. At this point it is helpful to reflect on the triumph of the Apple iPod again. Jonathan Ive designed the MP3 player in the spirit of Dieter Rams, the most prominent advocate of objectively functional design in the wake of the Ulm School of Design. The iPod unmistakably quotes Braun design's "simplicity equals beauty" formula from the sixties. When software determines the value of a product, function no longer has any significance for the design. It is the common memory that counts, for trust comes from familiarity. Quoting design ideas from the past is becoming a precondition for the future. The new technologies need the "old" culture in order to secure our trust.

ROBOTIC LIVING

Design in context with information technology – "intelligence as utopia"

The technological revolution is unstoppable, design has been assigned new tasks. Software programmes have meanwhile left virtual reality. The second reality of data is taking concrete shape and form. In the guise of mobile robots, data is beginning to interact with our world of the first reality, the world of biology and physics. What has come to be taken for granted in factory production still takes us by surprise in a private context. But the robots have long since left the factories and are in the process of conquering the household. They take on social functions. They become artificial family members, for the

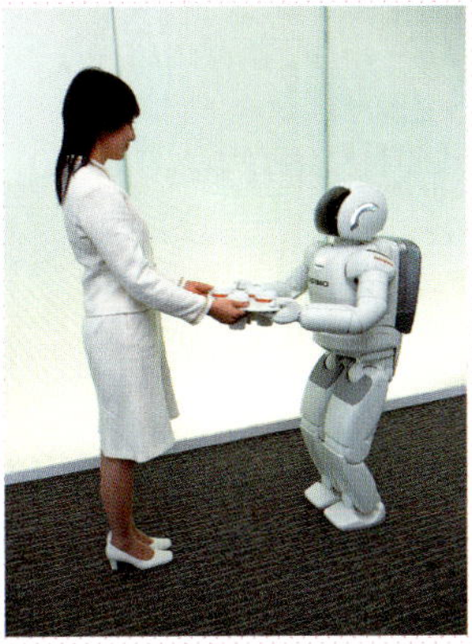

Honda deliberately equipped its "Asimo" with human proportions. Its height of 1.20 metre enables it to integrate with humans. It can operate door handles and light switches and work at conventional tables. The robot can also walk and go upstairs.

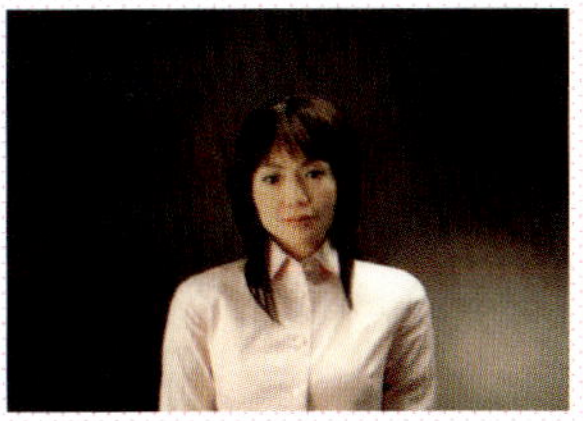

Hiroshi Ishiguro's "Osaka–Repliee Q1" humanoid is interactive and has the facial characteristics of an Asian woman. While robots with human figures trigger rejection in Europe, the Japanese take a far more pragmatic view of their usage.

moment still in the guise of programmed pets. The first computer-controlled dog goes by the name of "Aibo". He was presented by Sony in 1999; today third-generation models have entered the home. Aibos are mass-produced products with the ability to learn and, in the meantime, to communicate with the Internet. They have an MP3 player and audio e-mail and can adapt to their user's personal habits with ever-greater individuality. They are "born" without knowledge of their abilities and have to be "trained" by the owner.

The rationalisation of interpersonal affection by robots has already begun. The humanoid "Q1" is an informative robot with a striking similarity to an East Asian woman. When Q1 answers questions, "her" chest rises and falls. The skin-like finish gives her a very human appearance. Today we are on the threshold of a new era of interaction between man and machine. Whilst the robots are becoming increasingly "natural", prostheses for people are becoming increasingly artificial. Microprocessors control the hydraulics of the synthetic body parts and increase their efficiency. At the same time, a hyperrealistic plastic skin enhances their cultural acceptance. In the everyday life of those affected, Hollywood's Cyborg fantasies embody the hope of a "normal" life. With its artificial naturalness, robotic life is a new challenge for design.

NEXT NATURE

Design in context with bio- and nanotechnology – "paradise of the creators"

Tomorrow designers will have to prepare to make a quantum leap. Designing bio and nanotechnology will no longer be the exclusive domain of biologists and physicists. Technology philosopher Günther Anders realised at an early stage that "We are inverted utopians: While utopians cannot make what they imagine, we cannot imagine what we make." Headlines such as "healthy baby from embryo frozen 13 years ago" no longer cause a stir. In America these days, cloned pets only cost 32,000 US dollars. The animal-

Dutch designer Jurgen Bey covers mass-produced armchairs with PVC. The "Cocoon" collection (see p. 125) represents design's critical self-reflection on the threshold of the new challenges it faces.

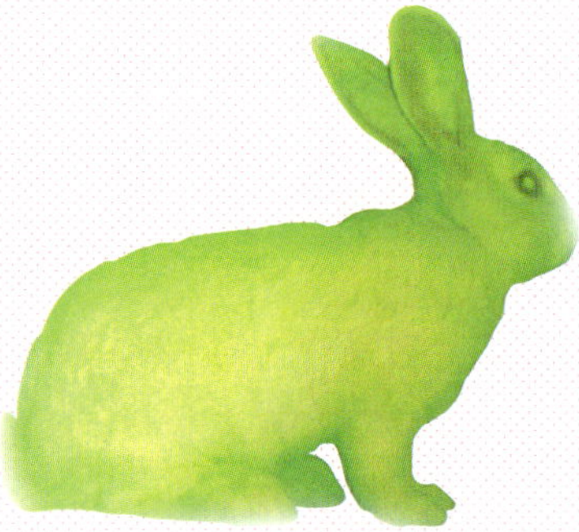

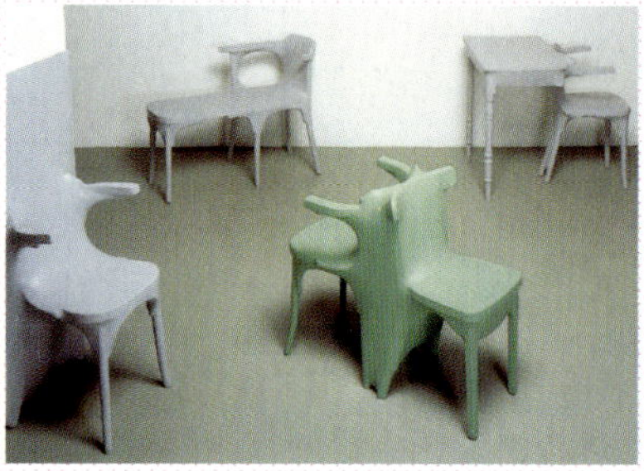

Brazilian artist Eduardo Kac links the reality of a gene laboratory with our everyday life in his "transgenic art". "Alba", a rabbit that glows green under UV light, created a scandal.

free production of meat, steak from regrown cells, will probably reach supermarket meat counters within the next 20 years. Biotechnology is turning meat, the substance we humans are made of, into a material of the future.

Nanotechnology is reorganising the natural laws of physics, chemistry or biology at atomic level. It is an umbrella term for a multitude of applications and products. Nanoparticles are so tiny that they are imperceptible to the naked eye. A human hair would have to be split 80,000 times to make it the same size as a nanometre. The bottom-up approach of nanotechnology involves building larger structures atom by atom and molecule by molecule or letting them grow as a result of self-organisation. Since nanoparticles have almost unlimited access to the human body, this technology will encounter a lack of trust. Neither the skin, the lungs nor the digestive tract are an obstacle to their mobility. Nor does the blood-brain barrier represent any major problem for the particles. Designers will be called upon to give this material a trustworthy shape and form. Tomorrow collaboration between biologists, electrical engineers and designers, a hitherto inconceivable proposition, will be something we take for granted.

It looks as if we are entering a magic garden that, while it amazes and enchants us, frightens and threatens us as well. We are in the process of redefining the notion of nature and have yet to design an appropriate symbolic form for the unknown things of this altered naturalness. The myth of creation is turning into a challenge. Will leaving our familiar environment lead to renewed expulsion or our return to paradise?

BANG – A FICTION?

In Michel Houellebecq's novel *The Possibility of an Island*, the clones are superhumans – in Kazuo Ishiguro's *Never Let Me Go* we see reality through the eyes of a clone – neither of them are science fiction stories, Ishiguro's novel takes place in England towards

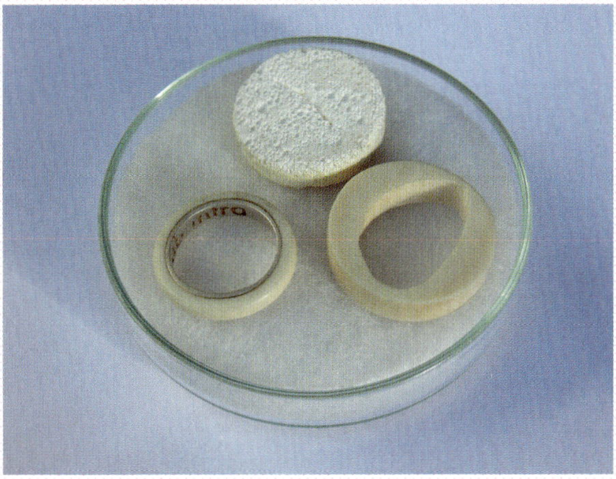

Designers Tobie Kerridge and Nikki Stott collaborated with bioengineer Ian Thompson to invent "Biojewellery". Ten couples participated and had wedding rings grown from bone marrow. Mounted on a ceramic frame (above), human tissue grows into bone structure (bottom right) which, when combined with silver, can be turned into jewellery. The picture depicts a design phase.

the end of the 20th century. – Stelios Arcadiou has grown himself a third ear. It is a ¼ scale replica of his left ear, grown from human cells in a bioreactor. Arcadiou calls himself Stelarc. – Eduardo Kac exhibits a green fluorescent rabbit. – Artist Paul McCarthy creates animated silicon pigs with quivering trotters and snouts. – In an exhibition of contemporary Chinese art at Bern Museum of Art, Xiao Yu presents *Ruan*, the head of a human foetus grafted onto an animal's body with the wings of a seagull.

What has hitherto populated science fiction scenarios in the form of mutants or paintings in the form of human-animal chimeras is leaving the realm of pure fantasy and becoming "real".

A NEW CONCEPTION OF DESIGN

A series of new terms is demonstrating the emergence of a new conception of design: Eugene Thacker entitles a book *Creative Biotechnology*, and in *Biomedia* describes the merging of information sciences and molecular biology, which Thacker contemplates from the perspective of a "biotech hobbyist", "biologics" are customised biological products and stem cell researchers refer to their work as "creating stem cells", designer drugs are, in a very literal sense, already on everybody's lips, German researcher Günter von Kiedrowski has constructed the prototype of a nano-scale self-reproducing object made of biological material, and designers like Nikki Stott and Tobie Kerridge have bred human bone cells into a ring, thus creating a truly individual piece of jewellery. The world can be reconstructed atom by atom, soon we will be able to dispense entirely with raw materials and naturally grown foods, and even today the taste of many products is no longer determined by strawberries or apples but by synthetically produced flavourings. However, this means the boundaries between "natural" and "artificial" that have determined

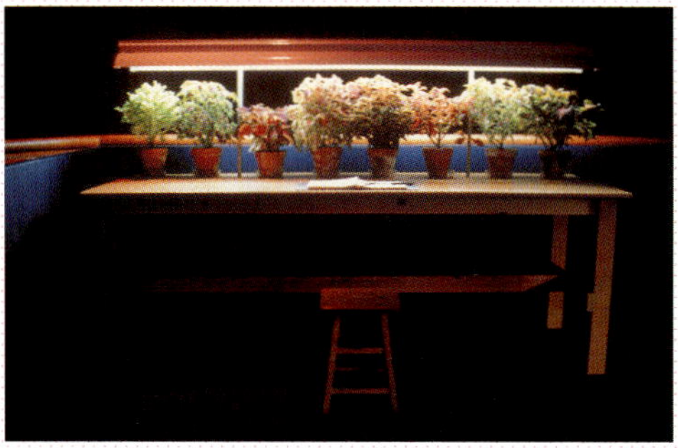

American artist George Gessert is interested in traditional breeding and selection processes. Genetics underscore the commonalities of all life forms. His "life-installations" place plants and animals within an art context.

our lives so far are disintegrating. Natural foods are turning into artificial foods. What we have hitherto called nature is becoming an environment designed by architects, with plants genetically adapted to it, technology is becoming biological, human biology is becoming technological – thanks to information and biotechnology. The notion of culture, which also encompasses design, is thus losing its opposition to nature. "Next nature", as Koert van Mensvoort calls it, is becoming part of our culture. Pat Mooney, winner of the Alternative Nobel Prize, paints a striking scenario of an imminent BANG future: "For the first time in history, God has a competitor capable of creating new living things and new substances." ENTRYPARADISE understands the striving of design and architecture as the striving for an ideal world, a new paradise. Society's willingness to take cultural risks is growing. The technological dimensions are increasingly assessed as controllable. The possibility of total control over our physical and biological environment is extolled as a return to paradise. The myth of creation is turning into a challenge. Life itself is increasingly becoming an aesthetic decision.

When human beings feel called upon as divine creators, the question as to design's responsibility will arise again. Society will have to decide how much "life design" it wants to permit – and must permit, in order to survive. For the dynamics of the new, all-embracing design do not only lead to paradise, but into the consciousness of a real risk society as well. How much risk a society considers reasonable for itself could well be the new value which the winners or losers of the globalisation debate will be measured by.

UP UNTIL NOW, DESIGNERS HAVE TRIED TO RECOGNISE A THING'S ESSENCE AND DERIVE AN APPROPRIATE FORM FROM IT. BACK IN 1896, AMERICAN ARCHITECT LOUIS SULLIVAN INSISTED THAT "LIFE IS RECOGNISABLE IN ITS EXPRESSION, THAT FORM EVER FOLLOWS FUNCTION". IN FUTURE, ADVANCING INTO NEW DIMENSIONS OF LIFE WILL PERMIT A NEW KIND OF FUNCTIONALISM THAT STARTS AT THE LEVEL OF ITS SMALLEST COMPONENT. THE OBJECT OF DESIGN WILL CHANGE ALONG WITH OUR UNDERSTANDING OF NATURE, THE METHODS AND GOALS OF DESIGN WILL BE RE-EVALUATED.

--- OBJECT AND NATURE ---

NORBERT BOLZ

BANG DESIGN

BANG design denotes the intersection today of two lines of development, namely the paradigm shift in design and the convergence of technologies. The acronym BANG stands for the basic components of our world: Bits, Atoms, Neurons and Genes. The fact that they must now be construed as a single unit is inextricably bound up with the magic dimension of ten to the power of minus nine meters. “Nano” denotes the scale of space between one and about 400 atoms, i.e., the field in which Newton’s physical laws are rendered inoperative by quantum behavior. At this nano-level, the difference between *natural* and *artificial* becomes obsolete as nanotechnology blurs the dividing line between natural and artificial molecular systems. Nanotechnology means that we can construct causal chains between the different scales.

Today, R&D for the first time focus on a material unity at the nano-level, and it is one at which almost all modern technologies can be integrated. It is therefore not overly audacious to expect a Golden Age, a new Renaissance that will herald the hierarchical integration of the sciences and a standardization of knowledge. The new picture of the world is one of a hierarchy of complex systems enracinated at the nano-level. While it is still difficult to distinguish here between science and fiction, between *hype* and *hope*. Yet anyone wishing not to miss out on this development, and it is being advanced in America with immense funding and great enthusiasm under the terse title of NBIC, will benefit less from skepticism and more from Newt Gingrich’s suggestion that we “be unreasonable”.

Essentially, the project of NBIC or BANG design hinges on four dreams man has: reaching understanding, creation, immortality and security.

BANG Design stands for the fundamental modules of our world: Bits, Atoms, Neurons, and Genes. Its core aim is to realize four of mankind's dreams: understanding, creation, immortality, and security. A paradigm shift led to design no longer being responsible for functional and factual transparence. The relationship between form and function in design deteriorated long ago, according to Norbert Bolz. It now has to offer security and world trust. Bolz explains that design's responsibility in the 21st century is to create a link between advanced nanotechnologies. This occurs on various levels. *Emotional design* designs emotional patterns and creates an outer skin for abstract feelings. *Black boxing* is the creation of user-friendly interfaces for complex systems. The digital world is complex and foreign to analog mankind. User-friendly in this context means the creation of functional simplicity within structural

I. Reaching understanding: The point here is the design of user-friendly interfaces, the construction of robots and personalized computers. The focus is on new communication-technologies and the opportunities for networking; on trying out designs in simulations and on new forms of learning.

II. Creation: Man's creativity quite literally becomes *Creation* when he creates not only as the architect of intelligent, user-friendly environments (intelligent houses, smart cities), but invents new materials that revolutionize clothing, living and mobility. And what we eat has long since ceased to be only produce from God's hand. New materials arise through the customized addition of molecules: *building* morphs into *growing*.

III. Immortality: *Forever young* is becoming a self-evident expectation for an ever growing number of people. In actual technical terms, the project *immortality* means that medicine and dietectics do everything (up to and including genetic engineering) to optimize the human body. Well-being becomes controllable through NBIC technologies and BANG design.

IV. Security: The greatest advances in technologies have got us no closer to eternal peace. At the same time, Western culture is less willing than ever to sacrifice human life in the pursuit of peace. Therefore, on the one hand the military has to deploy combat robots and unmanned strike planes or tanks, and, on the other, re-outfit modern warriors as almost inviolable cyborgs. Moreover, our security is increasingly threatened by asymmetrical enemies, by bio- and eco-terrorism, something that calls for surveillance measures on an unprecedented scale.

complexity for users. They accept not knowing what's going on inside the black box, as long as they can operate their computer with a user-friendly interface. "We have to understand interface design as the rhetoric of media technology," says Bolz. Adding challengingly, "if necessary, our civilization could dispense with intelligent human beings, but not thinking objects." A computer is responsible for more and more things that fill our world. And more and more things are products of design, like "food design", "body design", and even "designer babies". Design is becoming more important and more abstract. Today's co-evolution of technology and society is leading to socially intelligent and friendly technologies, to "personal machines". There are already computers that can be worn on or inside the body. Computers are even beginning to lead a life of their own. We are surrounded by relationship artifacts like

The focus in all these fields is on technically implemented knowledge. The interdisciplinary nature of scientific research in nano, bio, info, cogno and socio (i.e., physics, biology, mathematics/IT, brain research and the social sciences) is realized through the convergence of the leading technologies. Managing this convergence is the task design faces in the 21st century. The *language* of design creates and describes what it expresses; it embodies practical experiences and human expectations. One could also say: the design process is the virtual reality of practical experience.

In order to understand the outstanding importance of design for business and culture in the 21st century and justify the claim that the paradigm shift in design today results in the convergence of technologies, it helps to cast a brief glance at its history and reconstruct the first attempts to systematize a science of design.

The first design movement emerged in the struggle between three forces:
– first against the Empire style, i.e., the merely decorative, drapes, the devaluation of ornaments by their inflationary use, the dissolution of a room by the independent status gradually accorded furniture;
– second against the technology of mechanization such as, for example, the Jacquard loom, which led to a weakening of an awareness of materials;
– third against the emergence of mass phenomena and mass products in the arena of history.

Henry Cole championed a morality of pure forms through the application of the beautiful to mechanical production – *select pure forms* was his battle cry. One could say that Cole invented the modern art-&-craftsman. In a similar vein, William Morris then preached artistic morality. Materials and finishing, so he desired, were to again be appre-

tamogotchies or cuddly robots. BANG designers have a paradoxical task. They are working on new concept of nature called biodesign – making the designing of nature conceivable – or in other words, on a "posthuman paradise".

ciated as aesthetic values, and handicrafts seemed to be the antidote to the deterioration in quality.

In other words, initially design constitutes itself in polemic opposition to the modernization of society. However, art historian Alfred Gotthold Meyer noted as early as 1907, the year the German Werkbund was founded, that "the ferments that define the position of the 19th century in the history of style are the achievements of modern technology." In the battle of the academy artist and his stylistic forms versus the engineer and his structural formula, it was Art Nouveau that first suggested a synthesis. It was the take-off moment. Design became free forming. Art Nouveau stands for the moment when design breaks out of the ivory tower of art.

The foundation of the German Werkbund then clearly marks the point where design joins ranks with industry – we could talk here of the hour when industrial design was born. Art reorients itself in the world of machines. The spirit of the community of craftsmen gives way to the precision of industrial norms and the science of design.

What links Bauhaus to Futurism and Constructivism is the keen awareness of mechanical functions, industrial structures and new materials. This technical aesthetic disparages all pale humanisms and unleashes the scope for technical construction. Instead of preaching abstinence from technology in the name of damaged nature it practices a man/machine synergy under the conditions of new media.

Thus a new beauty is now to be nurtured alongside the beauty of nature and that of art, namely the "technically beautiful", whose shapes emerge in the "confines between art and technology" as Max Bense put it. The designer was no longer the supra-ordinated artist, but the partner in the decision-making process for industrial production. Design becomes understood in cybernetic terms as the design process with feedback loops – characteristic buzzwords here are *systems design* and *modular arrangement*.

Design shapes life, i.e., it is a balancing act between things, humans, and the environment. Otl Aicher speaks expressly of "life design" and of "optimizing the shape of life". In other words, design no longer focuses on individual objects, but on systems. Structure, grid, pattern, network – these are the key concepts of a non-linear design process that concurs with modern conditions of life.

Let us put the basic anthropological assumptions of this design concept somewhat more acutely: It is not the limits of the human body, but the limits of my instruments that are the limits of my world. Man is not a natural being and therefore man first is a human in synergy with his technologies. However, in order to understand technical, organized objects, you have to study them in use. The designer's world is not one of artistic forms, but forms of life. The attention to use is supposed to render the language of the objects themselves audible.

The *survival of the fittest* applies specifically to design: that is purposive which remains and fits in the interplay of forms of use. That is the applied principle of selection through tried-and-true use. In other words, the effects are the causes of the evolution of form. Here, design theory becomes cybernetic: The cause-effect chain is replaced by the circuit.

The talk is always of technologization whenever someone wishes to characterize realizations bereft of reflection. Here, the spirit of mathematics reigns, because it is largely procedural, i.e., it thinks for itself. And precisely in this spirit, people today are working to create intelligent technologies which, as the jargon of US high-tech institutes would have it, are *things that think*. An ever increasing amount of the thinking that has to be done is handled by things. Technology functions without consensus, and functioning technologies are something you cannot irritate.

Our civilization could at a punch thus get by without intelligent people, but not without thinking things. Progress in civilization is achieved not by what people think but what they are spared from having to think. Philosopher A.N. Whitehead saw this clearly decades ago when he suggested that civilization progresses exactly in that measure in which people can act without having to think about it. Thus, the social presumes technologization – against 'nature'. But this technology hides in the user's confidence.

The world of design was still in order as long as design followed self-evident use and form followed discernible function. However, in the age of microelectronics we find ourselves surrounded by black boxes to which there is no longer any intuitive access. Every household is familiar with the daily despair at the foreign language spoken by instructions for use. We have long since bidden farewell to the relationship to objects, nay the world, to what Heidegger with marvelous precision termed "Zuhandenheit". Something was "zuhanden" if it blended completely with its use and therefore never caught the eye as an object. Compare that with a video recorder.

The linkage of form and function has long since collapsed, and electronic miniaturization has caused objects to shrink to the size of flat user interfaces. Companies

Four historical, graphic key images that, in their time, described newly discovered phenomena in biology, physics, and technology were displayed as pictorial icons at the exhibition ENTRYPARADISE. The building blocks for a new manmade concept of creation were developed from hypotheses and models on bits, atoms, neurons, and genes.

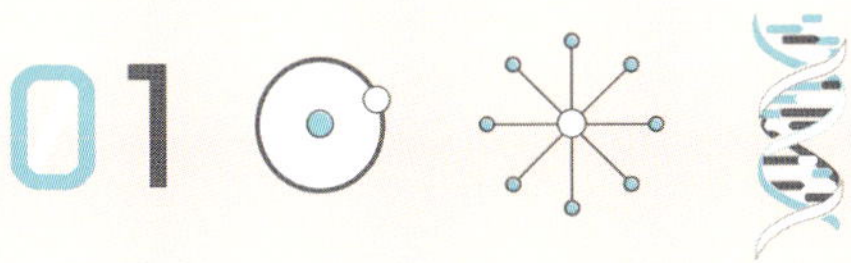

01

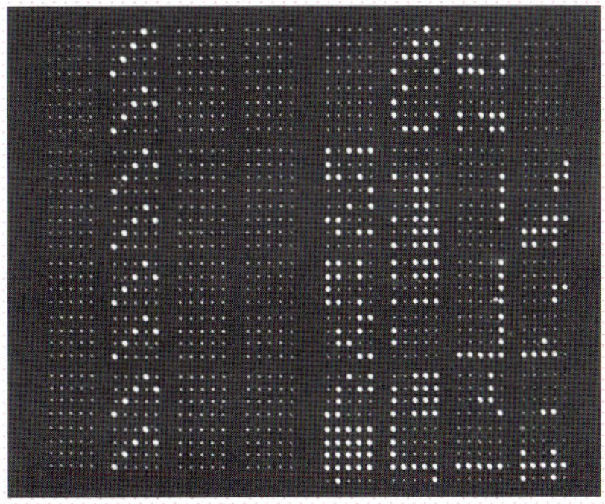

Bit: A cathode ray tube served as the main storage for the "Ferranti Mark 1", the computer that Frederic C. Williams and Tom Kilburn developed at Manchester University in 1949. Electrostatic-controlled phosphoric cells were used to store and display the individual data bits of 0 and 1. The Williams-Kilburn tube is considered the precursor of bitmap graphics.

manufacturing microelectronics, nanotechnology and biotech have one thing in common in particular: What they produce goes beyond the measure of man. Today, technology increasingly eludes visibility. Design can therefore no longer rely on functionalist/objective transparency, but must offer security and confidence in the world.

Emotional design focuses on showing people that a great deal is still possible, but at the same time protects them from the experience that everything is also possible in a different way, i.e., random. Design therefore no longer aims to grab the mind, but the immune system, the emotions. For it is not the mind but feelings that focus attention – they determine what is on the agenda. Precisely in recent years, sciences have recognized the emotions as a form of cognition and have been able to show that feelings correspond to behavioral patterns, i.e., in a sense have to be learned. And that is why it is also possible to modulate emotions. In the light of this, emotional design shapes patterns of feelings and thus creates a place for object-less feelings to tarry.

Intelligent design, self-explanatory use and *online help* are concepts that are destined to signal to the client: Forget technology! Design today no longer sees itself as the art of packaging a device, but as the animation of a black box. The latter kindles fear and it is the designer's primary task to allay our fears in this regard. Design must translate credibility into form. Should it succeed then the black box is a positive term and a knowledge of what goes on in it is not important for an understanding of its systemic function. And user is then an honorable name for the client of simplicity. He does not want to know anything about programs and processors, but to remain at the level of the friendly user interface.

And this of course impacts on our lifestyle. "We have learned to take things at interface value," Sherry Turkle rightly says. We accept it without knowing what happens in the black boxes. That functioned with cars for a long time, but today we are also seeing this

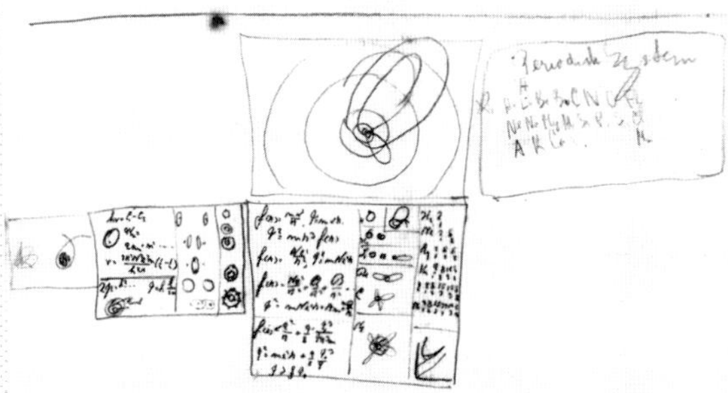

Atom: Research by Danish physicist Niels Bohr led in 1913 to a new atom model, where electrons travel in concentric orbit around a positively charged atom nucleus. A model which quickly proved too schematic. Bohr's sketches from 1921 already showed elliptical orbits of electrons; his model was later developed into today's typical orbital graphic imagery.

happen with regard to the computer as black box. For a knowledge of its circuits is not important in order to understand its social function. In other words, the question how a computer functions technically has nothing to do with the question what it means in social terms. You can understand the one without having any idea of the other.

Today, it is generally accepted that for design shaping the interface between telecommunication, the new media, and computer technologies will be one of the most important tasks of the future. Design science investigates the problems of a communications-centered use of technology. Is it a dot, a surface, a space, where users meet their gadgets, man meets machine? Where design with a twist of a magical wand allays our fears of impenetrable artifacts?

Needless to say, people have always created a world in which they can live. But it was not until an awareness of this as a problem arose that design arose in the emphatic sense. Designers want to access the world through design – instilling it with meaning. And if we today say goodbye to the design of things, then this orientational function of design becomes all the more clearly the center of attention. For this reason, object design is increasingly giving way to perception and lifestyle design. Just as science today does not start from the individual, but from relationships, postmodern design no longer starts from the object, but from a relationship.

However, even after the end of objectivity it is only those people who have a feel for design who have a sense of shape, i.e., understand the rhetoric of technology; and that in turn assumes an essentially positive relationship to the world of artifacts. Handling technology fosters a feel for shape. What I mean is the selection of the decorative effect, the mastery of the virtuoso, the joy of complex forms. In other words, design solves the problem of unity of complexity and familiarity. The designer wants to tempt us into using something and therefore has to work to eliminate people's fear of technology, the impenetrable gadget.

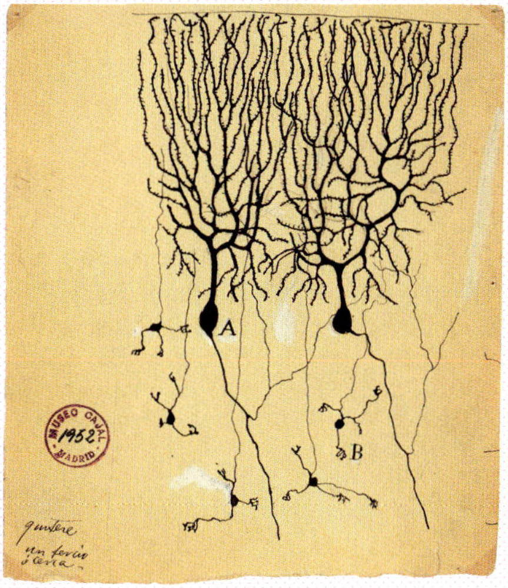

Neuron: Spanish doctor Santiago Ramón y Cajal researched and visualized the fine anatomy of nerve cells. Cajal's drawing from 1899 shows the Purkinje cells (A) and granule cells (B) of pigeon cerebellum. The researcher believed the nerve pathways of the adult brain to be rigid and unchanging. Today this is considered out-dated doctrine.

Anyone who means well with people protects them against digitality. Humans are analog beings – they want to have value and qualities, similarity and evidence. And the digital functions within humans have to date been well concealed by nature – namely our brain functions and the information of inheritance. Apparently, what happens in the central nervous system can easily be analogized (analogized!) using a computer's functional mode.

This is why a user-friendly computer makes sure I forget I am using a computer; its interface design shields me from the posthuman digital technology. Put succinctly: You can drive a car all your life without once having to look under the hood. And you can likewise work your entire life at a computer, without once having to look behind the user interface. In other words, user friendliness is exactly what sociologist Helmut Schelsky once termed "self-deception in terms of familiarity".

Put succinctly: user friendliness means functional simplicity with structural complexity – easy to use, hard to understand. The intelligence of products consists precisely of concealing the abyss of non-understanding. Thus, use detaches itself from understanding. Anyone talking today of intelligent design means that the use of an appliance is self-explanatory. However, this explanation does not lead to understanding, but to smooth functioning. One could therefore formulate it as follows: user friendliness is the rhetoric of technology that sanctifies our ignorance. And today this design-specific rhetoric which emerges precisely not in discourses but in the technology of interface design offers us the user illusion of world.

In other words, we must understand interface design as the rhetoric of media technology; and here dramatic changes have happened in the course of a few decades. As we all know, the history of the computer interface started with the punchcards used by IBM employees. The next step in the development of user interfaces was created by the

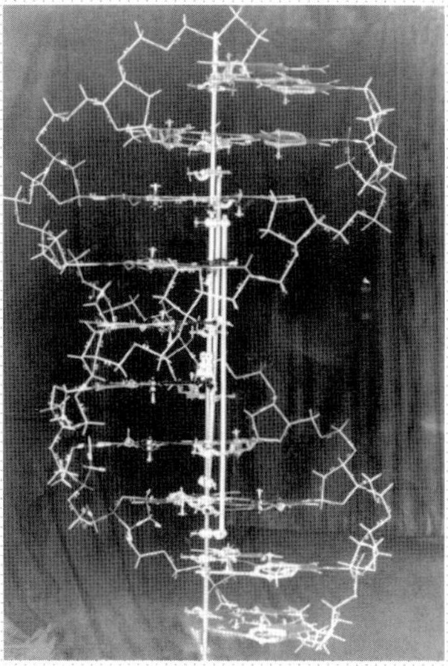

Gene: American biochemists and physicists James Watson and Francis Crick decoded the structure of deoxyribonucleic acid (DNA) in 1953. Their demonstration model shows the characteristic double helix structure, a long chain-molecule with the base pairs from which genes are composed.

alphanumeric command changes of the DOS Wizards – a new arcane science. Douglas Engelbart's Mouse and Apple's Macintosh made it possible to utilize programs by simply clicking icons. And the desktop metaphor then brought the major change in user friendliness. The next step in the evolution of interface design will lead into the virtual reality of cyberspace. In cyberspace, we can finally use computers by "natural" gestures – we no longer need to write programs but move in a vivid data space.

Historically speaking, design took its cue from the paradigms of decoration, function, emotion and then evolution. The first paradigm shift could be boiled down to the formula *form follows function*. After the end of Functionalism and New Objectivity, this changed into *form follows emotion*. And today the self-understanding of design as inspired by the theory of evolution must be *form follows form*; "forms taken out of the form", as logician George Spencer Brown puts it.

There is a computer in an increasing number of the things that surround us. And an increasing number of things which we encounter are the products of design. Ubiquitous computing – ubiquitous design. We quite naturally speak of designer baby, designer body, food design, etc. The technologies disappear in the fabric of everyday life. Design becomes ever more important – and abstract. This is why we look for attractive abstractions.

In the 19th century, metaphysics ended with the question as to the status of technology. Martin Heidegger expressly raised this question on technology and we must put it again today. A good starting point would be to remember that Aristotle defined the hand as the organ of all organs – and Alan Turing defined the computer as the machine of all machines. So it would seem obvious to grasp modern technology as a projection of organs and this brings us back into the lineage of Kapp, Freud and Marshall McLuhan. McLuhan's "extensions of man" and Freud's concept of the prosthetic god seem to fit particularly well into the world of "converging technologies".

The acronyms BANG and NBIC lack any reference to culture and society. But we must presume a co-evolution of technological and social systems, i.e., the field of tension between nano – bio – info – cogno – socio. The focus is thus on structures made with the modules of atoms – genes – bits – neurons – memes. The modules used by the social sciences still require getting used to: memes. One of the generative social sciences operating with these modules replaces the explanation of social phenomena with their computer simulation. Instead of asking *Can you explain it?* it wants to know *Can you grow it?*

The co-evolution of technology and society leads to a) socially intelligent and b) sociable technologies, i.e., personal machines. Today there are already computers we wear on or in our bodies. And technology is closing in on us, such as with the move from peripheral to implantable "devices". Moreover, computers are gaining a "kind of life", for example as robots, i.e., they are increasingly social actors. And in line with this users are developing a form of social behavior toward technologies. We can trace a line of relationship artifacts from toys and dolls, via Tamagochis and Furbies, through to the new teddy robots. Here, the focus is on designing artificial *subjectivities* with which you can develop a relationship. Technology that you like – that is the task of design today.

Scientists wishing to take part in BANG design have a paradoxical task: They must drive out not only *spirit* from the humanities but also *nature* from the natural sciences. Put differently they are working at a new concept of nature, namely biodesign, i.e., at something that makes the design of nature conceivable. Anyone seeking a metaphor for this task could talk of a new, posthuman paradise.

MONOGRAPHS

Dotzler, Bernhard; Kittler, Friedrich (eds.): *Turing, Alan: Intelligence Service – Ausgewählte Schriften* (Berlin, 1987)

Doyle, Richard M., *On Beyond Living. Rhetorical Transformations of the Life Sciences* (Stanford, CA. Stanford University Press, 1997)

Günther, Gotthard: *Beiträge zur Grundlegung einer operationsfähigen Dialektik*, (Hamburg, 3 vols., 1976–1980)

Johnson, Steven: *Everything Bad is Good for You: How Today's Popular Culture is Actually Making us Smarter* (London, 2005)

Johnson-Laird, Philip N.: *The Computer and the Mind: An Introduction to Cognitive Science*, (London, 1988)

Kurzweil, Ray: *The Age of the Spiritual Machines* (New York, 1999)

Lem, Stanislaw: *Die Vergangenheit der Zukunft* (Frankfurt, 1992)

Lem, Stanislaw: *Waffensysteme des 21. Jahrhunderts oder The Upside Down Evolution*, (Frankfurt, 1968)

Lupton, Ellen: *Skin: Surface, Substance and Design* (New York, 2002)

Markl, Hubert: *Evolution, Genetik und menschliches Verhalten* (Munich & Zurich, 1986)

Minsky, Marvin: *The Society of Mind* (New York, 1984)

Negroponte, Nicholas: *Being Digital* (1995)

Roco, Mihail C.; Bainbridge William Sims (eds.): *Societal Implications of Nanoscience and Nanotechnology* (Arlington, 2001)

Roco, Mihail C.; Bainbridge, William Sims (eds.): *Converging Technologies for Improving Human Performances* (Arlington, 2002)
Simon, Herbert A.: *The Sciences of the Artificial*, 3rd ed. (Cambridge, MA, 1996)
Turkle, Sherry: *Life of the Screen: Identity in the Age of the Internet* (New York, 1995)
Wackenroder, Wilhelm Heinrich, *Phantasien über die Kunst für Freunde der Kunst* (Hamburg, 1799)
Zons, Raimar: *Die Zeit des Menschen: Zur Kritik des Posthumanismus* (Frankfurt, 2001)

ARTICLES/ESSAYS IN COLLECTIONS
van den Boom, Holger: "Künstliche Intelligenz und Fiktion," in: *Strategien des Scheins, Kunst – Computer – Medien*, eds. Florian Rötzer & Peter Weibel (Munich, 1991), pp. 96–109
Metzger, Hans-Joachim: "Genesis in Silico," in: *HyperKult. Geschichte, Theorie und Kontext digitaler Medien*, eds. Martin Warnke, Wolfgang Coy & Georg Christoph Tholen (Basel/Frankfurt/Main, 1997), pp. 461–510

ARTICLES IN JOURNALS
"Bodybuilding: The Bionic Human," in: *Science, Special Issue*, Vol. 295, no. 5557 (Washington DC, 2002), pp. 995–1033
Joy, Bill: "Why the Future Doesn't Need Us," in: *Wired*, 8.04, April 2000 (also: www.wired.com/wired/archive/8.04/joy.html)

WEB SITES
www.educationarcade.org --- www.etcgroup.org ---
www.itri.loyola.edu/convergingTechnologies --- www.wtec.org

BILL JOY

WHY THE FUTURE DOESN'T NEED US

From the moment I became involved in the creation of new technologies, their ethical dimensions have concerned me, but it was only in the autumn of 1998 that I became anxiously aware of how great are the dangers facing us in the 21st century. I can date the onset of my unease to the day I met Ray Kurzweil, the deservedly famous inventor of the first reading machine for the blind and many other amazing things.

Ray and I were both speakers at George Gilder's Telecosm conference, and I encountered him by chance in the bar of the hotel after both our sessions were over. I was sitting with John Searle, a Berkeley philosopher who studies consciousness. While we were talking, Ray approached and a conversation began, the subject of which haunts me to this day.

I had missed Ray's talk and the subsequent panel that Ray and John had been on, and they now picked right up where they'd left off, with Ray saying that the rate of improvement of technology was going to accelerate and that we were going to become robots or fuse with robots or something like that, and John countering that this couldn't happen, because the robots couldn't be conscious.

While I had heard such talk before, I had always felt sentient robots were in the realm of science fiction. But now, from someone I respected, I was hearing a strong argument that they were a near-term possibility. I was taken aback, especially given Ray's proven ability to imagine and create the future. I already knew that new technologies like genetic engineering and nanotechnology were giving us the power to remake the world, but a realistic and imminent scenario for intelligent robots surprised me. It's easy to get jaded about such breakthroughs. We hear in the news almost every day of some kind of

Bill Joy's text, written in 2000, is a caution against new technologies like robotics, gene technology, and nanotechnologies. He stated, one year previous to the September 11, that they will increase the danger of terrorist attacks. Without evoking a horror scenario, the computer specialist immediately triggered a wide debate among those for and against his theories. He points out the concrete dangers that could result from the momentum of progressive research and development. Even individuals and small groups would be able to abuse 21st century technologies for military or terrorist means. You only need knowledge, not large systems or rare raw materials, to destroy certain areas or people with specific genetic characteristics. Joy hypothesizes that we will be able to produce machines in 2030 one million times more efficient than today's personal computer. This includes robots and nanobots

technological or scientific advance. Yet this was no ordinary prediction. In the hotel bar, Ray gave me a partial preprint of his then-forthcoming book *The Age of Spiritual Machines*, which outlined a utopia he foresaw – one in which humans gained near immortality by becoming one with robotic technology. On reading it, my sense of unease only intensified; I felt sure he had to be understating the dangers, understating the probability of a bad outcome along this path.

I found myself most troubled by a passage detailing a *dys*topian scenario:

First let us postulate that the computer scientists succeed in developing intelligent machines that can do all things better than human beings can do them. In that case presumably all work will be done by vast, highly organized systems of machines and no human effort will be necessary. Either of two cases might occur. The machines might be permitted to make all of their own decisions without human oversight, or else human control over the machines might be retained. If the machines are permitted to make all their own decisions, we can't make any conjectures as to the results, because it is impossible to guess how such machines might behave.

We only point out that the fate of the human race would be at the mercy of the machines. It might be argued that the human race would never be foolish enough to hand over all the power to the machines. But we are suggesting neither that the human race would voluntarily turn power over to the machines nor that the machines would willfully seize power. What we do suggest is that the human race might easily permit itself to drift into a position of such dependence on the machines that it would have no practical choice but to accept all of the machines' decisions. As society and the problems that face it become more and more complex and machines become more and more intelligent, people will let machines make more of their decisions for them, simply because machine-made decisions will bring better results than man-made ones. Eventually a stage may be reached

with complex intelligence and automatic control mechanisms. According to Joy, it is only one small step before they are able to copy themselves and annihilate mankind. Joy bases his theory on texts by the disturbed Unabomber Theodore Kaczynski, on writings by robotics researcher Hans Moravec, and on George Dyson, author of *Darwinism Among the Machines*. Joy argues that we have to relinquish developing technologies that are too dangerous. Gene technology, robotics, and nanotechnologies are difficult to separate from civil and military application. Yet forbidding these is not a realistic alternative. It would require gigantic surveillance systems and restrict mankind's privacy. Joy calls for specialists in new technology to voluntarily renounce this research and development.

at which the decisions necessary to keep the system running will be so complex that human beings will be incapable of making them intelligently. At that stage the machines will be in effective control. People won't be able to just turn the machines off, because they will be so dependent on them that turning them off would amount to suicide.

On the other hand it is possible that human control over the machines may be retained. In that case the average man may have control over certain private machines of his own, such as his car or his personal computer, but control over large systems of machines will be in the hands of a tiny elite – just as it is today, but with two differences. Due to improved techniques the elite will have greater control over the masses; and because human work will no longer be necessary, the masses will be superfluous, a useless burden on the system. If the elite is ruthless they may simply decide to exterminate the mass of humanity. If they are humane they may use propaganda or other psychological or biological techniques to reduce the birth rate until the mass of humanity becomes extinct, leaving the world to the elite. Or, if the elite consists of soft-hearted liberals, they may decide to play the role of good shepherds to the rest of the human race. They will see to it that everyone's physical needs are satisfied, that all children are raised under psychologically hygienic conditions, that everyone has a wholesome hobby to keep him busy, and that anyone who may become dissatisfied undergoes "treatment" to cure his "problem." Of course, life will be so purposeless that people will have to be biologically or psychologically engineered either to remove their need for the power process or make them "sublimate" their drive for power into some harmless hobby. These engineered human beings may be happy in such a society, but they will most certainly not be free. They will have been reduced to the status of domestic animals.

In the book, you don't discover until you turn the page that the author of this passage is Theodore Kaczynski – the Unabomber. I am no apologist for Kaczynski. His bombs

I started showing friends the Kaczynski quote; I would hand them the book, let them read the quote, and then watch their reaction as they discovered who had written it: the Unabomber.

killed three people during a 17-year terror campaign and wounded many others. One of his bombs gravely injured my friend David Gelernter, one of the most brilliant and visionary computer scientists of our time. Like many of my colleagues, I felt that I could easily have been the Unabomber's next target.

Kaczynski's actions were murderous and, in my view, criminally insane. He is clearly a Luddite, but simply saying this does not dismiss his argument; as difficult as it is for me to acknowledge, I saw some merit in the reasoning in this single passage. I felt compelled to confront it.

Kaczynski's dystopian vision describes unintended consequences, a well-known problem with the design and use of technology, and one that is clearly related to Murphy's law – "Anything that can go wrong, will." (Actually, this is Finagle's law, which in itself shows that Finagle was right.) Our overuse of antibiotics has led to what may be the biggest such problem so far: the emergence of antibiotic-resistant and much more dangerous bacteria. Similar things happened when attempts to eliminate malarial mosquitoes using DDT caused them to acquire DDT resistance; malarial parasites likewise acquired multi-drug-resistant genes.

The cause of many such surprises seems clear: The systems involved are complex, involving interaction among and feedback between many parts. Any changes to such a system will cascade in ways that are difficult to predict; this is especially true when human actions are involved.

I started showing friends the Kaczynski quote from *The Age of Spiritual Machines*; I would hand them Kurzweil's book, let them read the quote, and then watch their reaction as they discovered who had written it. At around the same time, I found Hans Moravec's book *Robot: Mere Machine to Transcendent Mind*. Moravec is one of the leaders in robotics research, and was a founder of the world's largest robotics research program, at

Carnegie Mellon University. *Robot* gave me more material to try out on my friends – material surprisingly supportive of Kaczynski's argument.

For example: Biological species almost never survive encounters with superior competitors. Ten million years ago, South and North America were separated by a sunken Panama isthmus. South America, like Australia today, was populated by marsupial mammals, including pouched equivalents of rats, deers, and tigers. When the isthmus connecting North and South America rose, it took only a few thousand years for the northern placental species, with slightly more effective metabolisms and reproductive and nervous systems, to displace and eliminate almost all the southern marsupials.

In a completely free marketplace, superior robots would surely affect humans as North American placentals affected South American marsupials (and as humans have affected countless species). Robotic industries would compete vigorously among themselves for matter, energy, and space, incidentally driving their price beyond human reach. Unable to afford the necessities of life, biological humans would be squeezed out of existence.

There is probably some breathing room, because we do not live in a completely free marketplace. Government coerces nonmarket behavior, especially by collecting taxes. Judiciously applied, governmental coercion could support human populations in high style on the fruits of robot labor, perhaps for a long while.

A textbook dystopia – and Moravec is just getting wound up. He goes on to discuss how our main job in the 21st century will be "ensuring continued cooperation from the robot industries" by passing laws decreeing that they be "nice," and to describe how seriously dangerous a human can be "once transformed into an unbounded superintelligent robot." Moravec's view is that the robots will eventually succeed us – that humans clearly face extinction.

I decided it was time to talk to my friend Danny Hillis. Danny became famous as the cofounder of Thinking Machines Corporation, which built a very powerful parallel supercomputer. Despite my current job title of Chief Scientist at Sun Microsystems, I am more a computer architect than a scientist, and I respect Danny's knowledge of the information and physical sciences more than that of any other single person I know. Danny is also a highly regarded futurist who thinks long-term – four years ago he started the Long Now Foundation, which is building a clock designed to last 10,000 years, in an attempt to draw attention to the pitifully short attention span of our society.

So I flew to Los Angeles for the express purpose of having dinner with Danny and his wife, Pati. I went through my now-familiar routine, trotting out the ideas and passages that I found so disturbing. Danny's answer – directed specifically at Kurzweil's scenario of humans merging with robots – came swiftly, and quite surprised me. He said, simply, that the changes would come gradually, and that we would get used to them.

But I guess I wasn't totally surprised. I had seen a quote from Danny in Kurzweil's book in which he said, "I'm as fond of my body as anyone, but if I can be 200 with a body of silicon, I'll take it." It seemed that he was at peace with this process and its attendant risks, while I was not.

While talking and thinking about Kurzweil, Kaczynski, and Moravec, I suddenly remembered a novel I had read almost 20 years ago – *The White Plague*, by Frank Herbert – in which a molecular biologist is driven insane by the senseless murder of his family. To seek revenge he constructs and disseminates a new and highly contagious plague that kills widely but selectively. (We're lucky Kaczynski was a mathematician, not a molecular biologist.) I was also reminded of the Borg of *Star Trek*, a hive of partly biological, partly robotic creatures with a strong destructive streak. Borg-like disasters are a staple of science fiction, so why hadn't I been more concerned about such robotic dystopias earlier? Why weren't other people more concerned about these nightmarish scenarios?

Part of the answer certainly lies in our attitude toward the new – in our bias toward instant familiarity and unquestioning acceptance. Accustomed to living with almost routine scientific breakthroughs, we have yet to come to terms with the fact that the most compelling 21st-century technologies – robotics, genetic engineering, and nanotechnology – pose a different threat than the technologies that have come before. Specifically, robots, engineered organisms, and nanobots share a dangerous amplifying factor: They can self-replicate. A bomb is blown up only once – but one bot can become many, and quickly get out of control.

Much of my work over the past 25 years has been on computer networking, where the sending and receiving of messages creates the opportunity for out-of-control replication. But while replication in a computer or a computer network can be a nuisance, at worst it disables a machine or takes down a network or network service. Uncontrolled self-replication in these newer technologies runs a much greater risk: a risk of substantial damage in the physical world.

Each of these technologies also offers untold promise: The vision of near immortality that Kurzweil sees in his robot dreams drives us forward; genetic engineering may soon provide treatments, if not outright cures, for most diseases; and nanotechnology and nanomedicine can address yet more ills. Together they could significantly extend our average life span and improve the quality of our lives. Yet, with each of these technologies, a sequence of small, individually sensible advances leads to an accumulation of great power and, concomitantly, great danger.

What was different in the 20th century? Certainly, the technologies underlying the weapons of mass destruction (WMD) – nuclear, biological, and chemical (NBC) – were powerful, and the weapons an enormous threat. But building nuclear weapons required, at least for a time, access to both rare – indeed, effectively unavailable – raw materials and highly protected information; biological and chemical weapons programs also tended to require large-scale activities.

The 21st-century technologies – genetics, nanotechnology, and robotics (GNR) – are so powerful that they can spawn whole new classes of accidents and abuses. Most dangerously, for the first time, these accidents and abuses are widely within the reach of individuals or small groups. They will not require large facilities or rare raw materials. Knowledge alone will enable the use of them.

Thus we have the possibility not just of weapons of mass destruction but of knowledge-enabled mass destruction (KMD), this destructiveness hugely amplified by the power of self-replication.

I think it is no exaggeration to say we are on the cusp of the further perfection of extreme evil, an evil whose possibility spreads well beyond that which weapons of mass destruction bequeathed to the nation-states, on to a surprising and terrible empowerment of extreme individuals.

Nothing about the way I got involved with computers suggested to me that I was going to be facing these kinds of issues.

My life has been driven by a deep need to ask questions and find answers. When I was three, I was already reading, so my father took me to the elementary school, where I sat on the principal's lap and read him a story. I started school early, later skipped a grade, and escaped into books – I was incredibly motivated to learn. I asked lots of questions, often driving adults to distraction.

As a teenager I was very interested in science and technology. I wanted to be a ham radio operator but didn't have the money to buy the equipment. Ham radio was the Internet of its time: very addictive, and quite solitary. Money issues aside, my mother put her foot down – I was not to be a ham; I was antisocial enough already.

I may not have had many close friends, but I was awash in ideas. By high school, I had discovered the great science fiction writers. I remember especially Heinlein's *Have Spacesuit Will Travel* and Asimov's *I, Robot*, with its Three Laws of Robotics. I was enchanted by the descriptions of space travel, and wanted to have a telescope to look at the stars; since I had no money to buy or make one, I checked books on telescope-making out of the library and read about making them instead. I soared in my imagination.

Thursday nights my parents went bowling, and we kids stayed home alone. It was the night of Gene Roddenberry's original *Star Trek*, and the program made a big impression on me. I came to accept its notion that humans had a future in space, Western-style, with big heroes and adventures. Roddenberry's vision of the centuries to come was one with strong moral values, embodied in codes like the Prime Directive: to not interfere in the development of less technologically advanced civilizations. This had an incredible appeal to me; ethical humans, not robots, dominated this future, and I took Roddenberry's dream as part of my own.

I excelled in mathematics in high school, and when I went to the University of Michigan as an undergraduate engineering student I took the advanced curriculum of the mathematics majors. Solving math problems was an exciting challenge, but when I discovered computers I found something much more interesting: a machine into which you could put a program that attempted to solve a problem, after which the machine quickly checked the solution. The computer had a clear notion of correct and incorrect, true and false. Were my ideas correct? The machine could tell me. This was very seductive.

I was lucky enough to get a job programming early supercomputers and discovered the amazing power of large machines to numerically simulate advanced designs. When I went to graduate school at UC Berkeley in the mid-1970s, I started staying up late, often all night, inventing new worlds inside the machines. Solving problems. Writing the code that argued so strongly to be written.

In *The Agony and the Ecstasy*, Irving Stone's biographical novel of Michelangelo, Stone described vividly how Michelangelo released the statues from the stone, "breaking the marble spell," carving from the images in his mind. In my most ecstatic moments, the software in the computer emerged in the same way. Once I had imagined it in my mind I felt that it was already there in the machine, waiting to be released. Staying up all night seemed a small price to pay to free it – to give the ideas concrete form.

Still, by the early 1980s, I was drowning in work. The Unix releases were very successful, and my little project of one soon had money and some staff, but the problem at Berkeley was always office space rather than money – there wasn't room for the help the project needed, so when the other founders of Sun Microsystems showed up I jumped at the chance to join them. At Sun, the long hours continued into the early days of workstations and personal computers, and I have enjoyed participating in the creation of advanced microprocessor technologies and Internet technologies such as Java and Jini.

From all this, I trust it is clear that I am not a Luddite. I have always, rather, had a strong belief in the value of the scientific search for truth and in the ability of great engineering to bring material progress. The Industrial Revolution has immeasurably improved everyone's life over the last couple hundred years, and I always expected my career to involve the building of worthwhile solutions to real problems, one problem at a time.

I have not been disappointed. My work has had more impact than I had ever hoped for and has been more widely used than I could have reasonably expected. I have spent the last 20 years still trying to figure out how to make computers as reliable as I want them to be (they are not nearly there yet) and how to make them simple to use (a goal that has met with even less relative success). Despite some progress, the problems that remain seem even more daunting.

But while I was aware of the moral dilemmas surrounding technology's consequences in fields like weapons research, I did not expect that I would confront such issues in my own field, or at least not so soon.

Perhaps it is always hard to see the bigger impact while you are in the vortex of a change. Failing to understand the consequences of our inventions while we are in the rapture of discovery and innovation seems to be a common fault of scientists and technologists; we have long been driven by the overarching desire to know that is the nature of science's quest, not stopping to notice that the progress to newer and more powerful technologies can take on a life of its own.

I have long realized that the big advances in information technology come not from the work of computer scientists, computer architects, or electrical engineers, but from that of physical scientists. The physicists Stephen Wolfram and Brosl Hasslacher introduced me, in the early 1980s, to chaos theory and nonlinear systems. In the 1990s, I learned about complex systems from conversations with Danny Hillis, the biologist Stuart Kauffman, the Nobel-laureate physicist Murray Gell-Mann, and others. Most recently, Hasslacher and the electrical engineer and device physicist Mark Reed have been giving me insight into the incredible possibilities of molecular electronics.

In my own work, as codesigner of three microprocessor architectures – SPARC, picoJava, and MAJC – and as the designer of several implementations thereof, I've been afforded a deep and firsthand acquaintance with Moore's law. For decades, Moore's law has correctly predicted the exponential rate of improvement of semiconductor technology. Until last year I believed that the rate of advances predicted by Moore's law might continue only until roughly 2010, when some physical limits would begin to be reached. It was not obvious to me that a new technology would arrive in time to keep performance advancing smoothly.

But because of the recent rapid and radical progress in molecular electronics – where individual atoms and molecules replace lithographically drawn transistors – and related nanoscale technologies, we should be able to meet or exceed the Moore's law rate of progress for another 30 years. By 2030, we are likely to be able to build machines, in quantity, a million times as powerful as the personal computers of today – sufficient to implement the dreams of Kurzweil and Moravec.

As this enormous computing power is combined with the manipulative advances of the physical sciences and the new, deep understandings in genetics, enormous transformative power is being unleashed. These combinations open up the opportunity to completely redesign the world, for better or worse: The replicating and evolving processes that have been confined to the natural world are about to become realms of human endeavor.

In designing software and microprocessors, I have never had the feeling that I was designing an intelligent machine. The software and hardware is so fragile and the capabilities of the machine to "think" so clearly absent that, even as a possibility, this has always seemed very far in the future.

But now, with the prospect of human-level computing power in about 30 years, a new idea suggests itself: that I may be working to create tools which will enable the construction of the technology that may replace our species. How do I feel about this? Very uncomfortable. Having struggled my entire career to build reliable software systems, it seems to me more than likely that this future will not work out as well as some people may imagine. My personal experience suggests we tend to overestimate our design abilities.

Given the incredible power of these new technologies, shouldn't we be asking how we can best coexist with them? And if our own extinction is a likely, or even possible, outcome of our technological development, shouldn't we proceed with great caution?

The dream of robotics is, first, that intelligent machines can do our work for us, allowing us lives of leisure, restoring us to Eden. Yet in his history of such ideas *Darwin Among the Machines*, George Dyson warns: "In the game of life and evolution there are three players at the table: human beings, nature, and machines. I am firmly on the side of nature. But nature, I suspect, is on the side of the machines." As we have seen, Moravec agrees, believing we may well not survive the encounter with the superior robot species.

How soon could such an intelligent robot be built? The coming advances in computing power seem to make it possible by 2030. And once an intelligent robot exists, it is only a small step to a robot species – to an intelligent robot that can make evolved copies of itself.

A second dream of robotics is that we will gradually replace ourselves with our robotic technology, achieving near immortality by downloading our consciousnesses; it is this process that Danny Hillis thinks we will gradually get used to and that Ray Kurzweil elegantly details in *The Age of Spiritual Machines*.

But if we are downloaded into our technology, what are the chances that we will thereafter be ourselves or even human? It seems to me far more likely that a robotic existence would not be like a human one in any sense that we understand, that the robots would in no sense be our children, that on this path our humanity may well be lost.

Genetic engineering promises to revolutionize agriculture by increasing crop yields while reducing the use of pesticides; to create tens of thousands of novel species of bacteria, plants, viruses, and animals; to replace reproduction, or supplement it, with cloning; to create cures for many diseases, increasing our life span and our quality of life; and much, much more. We now know with certainty that these profound changes in the biological sciences are imminent and will challenge all our notions of what life is.

Technologies such as human cloning have in particular raised our awareness of the profound ethical and moral issues we face. If, for example, we were to reengineer ourselves into several separate and unequal species using the power of genetic engineering, then we would threaten the notion of equality that is the very cornerstone of our democracy.

Given the incredible power of genetic engineering, it's no surprise that there are significant safety issues in its use. My friend Amory Lovins recently co-wrote, along with Hunter Lovins, an editorial that provides an ecological view of some of these dangers. Among their concerns: that "the new botany aligns the development of plants with their

But now a new idea suggests itself: that I may be working to create tools which will enable the construction of the technology that may replace our species.

economic, not evolutionary, success." Amory's long career has been focused on energy and resource efficiency by taking a whole-system view of human-made systems; such a whole-system view often finds simple, smart solutions to otherwise seemingly difficult problems, and is usefully applied here as well.

After reading the Lovins' editorial, I saw an op-ed by Gregg Easterbrook in *The New York Times* (November 19, 1999) about genetically engineered crops, under the headline: "Food for the Future: Someday, rice will have built-in vitamin A. Unless the Luddites win." Are Amory and Hunter Lovins Luddites? Certainly not. I believe we all would agree that golden rice, with its built-in vitamin A, is probably a good thing, if developed with proper care and respect for the likely dangers in moving genes across species boundaries. Awareness of the dangers inherent in genetic engineering is beginning to grow, as reflected in the Lovins' editorial. The general public is aware of, and uneasy about, genetically modified foods, and seems to be rejecting the notion that such foods should be permitted to be unlabeled.

But genetic engineering technology is already very far along. As the Lovins note, the USDA has already approved about 50 genetically engineered crops for unlimited release; more than half of the world's soybeans and a third of its corn now contain genes spliced-in from other forms of life.

While there are many important issues here, my own major concern with genetic engineering is narrower: that it gives the power – whether militarily, accidentally or in a deliberate terrorist act – to create a White Plague.

The many wonders of nanotechnology were first imagined by the Nobel-laureate physicist Richard Feynman in a speech he gave in 1959, subsequently published under the title "There's Plenty of Room at the Bottom." The book that made a big impression on me, in the mid-'80s, was Eric Drexler's *Engines of Creation*, in which he described beautifully how manipulation of matter at the atomic level could create a utopian future of abundance, where just about everything could be made cheaply, and almost any imaginable disease or physical problem could be solved using nanotechnology and artificial intelligences.

A subsequent book, *Unbounding the Future: The Nanotechnology Revolution*, which Drexler co-wrote, imagines some of the changes that might take place in a world where we had molecular-level "assemblers". Assemblers could make possible incredibly low-cost solar power, cures for cancer and the common cold by augmentation of the human immune system, essentially complete cleanup of the environment, incredibly inexpensive pocket supercomputers – in fact, any product would be manufacturable by assemblers at a cost no greater than that of wood – spaceflight more accessible than transoceanic travel today, and restoration of extinct species.

I remember feeling good about nanotechnology after reading *Engines of Creation*. As a technologist, it gave me a sense of calm – that is, nanotechnology showed us that incredible progress was possible, and indeed perhaps inevitable. If nanotechnology was our future, then I didn't feel pressed to solve so many problems in the present. I would get to Drexler's utopian future in due time; I might as well enjoy life more in the here and now. It didn't make sense, given his vision, to stay up all night, all the time.

Drexler's vision also led to a lot of good fun. I would occasionally get to describe the wonders of nanotechnology to others who had not heard of it. After teasing them with all the things Drexler described I would give a homework assignment of my own: "Use nanotechnology to create a vampire; for extra credit create an antidote."

With these wonders came clear dangers, of which I was acutely aware. As I said at a nanotechnology conference in 1989, "We can't simply do our science and not worry about these ethical issues." But my subsequent conversations with physicists convinced me that nanotechnology might not even work – or, at least, it wouldn't work anytime soon. Shortly thereafter I moved to Colorado, to a skunk works I had set up, and the focus of my work shifted to software for the Internet, specifically on ideas that became Java and Jini.

Then, last summer, Brosl Hasslacher told me that nanoscale molecular electronics was now practical. This was new news, at least to me, and I think to many people – and it radically changed my opinion about nanotechnology. It sent me back to *Engines of Creation*. Rereading Drexler's work after more than 10 years, I was dismayed to realize how little I had remembered of its lengthy section called "Dangers and Hopes", including a discussion of how nanotechnologies can become "engines of destruction". Indeed, in my rereading of this cautionary material today, I am struck by how naive some of Drexler's safeguard proposals seem, and how much greater I judge the dangers to be now than even he seemed to then. (Having anticipated and described many technical and political problems with nanotechnology, Drexler started the Foresight Institute in the late 1980s "to help prepare society for anticipated advanced technologies" – most important, nanotechnology.)

The enabling breakthrough to assemblers seems quite likely within the next 20 years. Molecular electronics – the new subfield of nanotechnology where individual molecules are circuit elements – should mature quickly and become enormously lucrative within this decade, causing a large incremental investment in all nanotechnologies.

Unfortunately, as with nuclear technology, it is far easier to create destructive uses for nanotechnology than constructive ones. Nanotechnology has clear military and terrorist uses, and you need not be suicidal to release a massively destructive nanotechnological device – such devices can be built to be selectively destructive, affecting, for example, only a certain geographical area or a group of people who are genetically distinct.
An immediate consequence of the Faustian bargain in obtaining the great power of nanotechnology is that we run a grave risk – the risk that we might destroy the biosphere on which all life depends.

In truth, we have had in hand for years clear warnings of the dangers inherent in widespread knowledge of GNR technologies – of the possibility of knowledge alone enabling mass destruction. But these warnings haven't been widely publicized; the public discussions have been clearly inadequate. There is no profit in publicizing the dangers.

The nuclear, biological, and chemical (NBC) technologies used in 20th-century weapons of mass destruction were and are largely military, developed in government laboratories. In sharp contrast, the 21st-century GNR technologies have clear commercial uses and are being developed almost exclusively by corporate enterprises. In this age of triumphant commercialism, technology – with science as its handmaiden – is delivering a series of almost magical inventions that are the most phenomenally lucrative ever seen. We are aggressively pursuing the promises of these new technologies within the now-unchallenged system of global capitalism and its manifold financial incentives and competitive pressures.

We should have learned a lesson from the making of the first atomic bomb and the resulting arms race. We didn't do well then, and the parallels to our current situation are troubling.

In our time, how much danger do we face, not just from nuclear weapons, but from all of these technologies? How high are the extinction risks?

The philosopher John Leslie has studied this question and concluded that the risk of human extinction is at least 30 percent, while Ray Kurzweil believes we have "a better than even chance of making it through," with the caveat that he has "always been accused of being an optimist." Not only are these estimates not encouraging, but they do not include the probability of many horrid outcomes that lie short of extinction.

Faced with such assessments, some serious people are already suggesting that we simply move beyond Earth as quickly as possible. We would colonize the galaxy using von Neumann probes, which hop from star system to star system, replicating as they go. This step will almost certainly be necessary five billion years from now (or sooner if our solar system is disastrously impacted by the impending collision of our galaxy with the An-

Many other people who know about the dangers still seem strangely silent.
I don't know where these people hide their fear.

dromeda galaxy within the next three billion years), but if we take Kurzweil and Moravec at their word it might be necessary by the middle of this century.

What are the moral implications here? If we must move beyond Earth this quickly in order for the species to survive, who accepts the responsibility for the fate of those (most of us, after all) who are left behind? And even if we scatter to the stars, isn't it likely that we may take our problems with us or find, later, that they have followed us? The fate of our species on Earth and our fate in the galaxy seem inextricably linked.

Another idea is to erect a series of shields to defend against each of the dangerous technologies. The Strategic Defense Initiative, proposed by the Reagan administration, was an attempt to design such a shield against the threat of a nuclear attack from the Soviet Union. But as Arthur C. Clarke, who was privy to discussions about the project, observed: "Though it might be possible, at vast expense, to construct local defense systems that would 'only' let through a few percent of ballistic missiles, the much touted idea of a national umbrella was nonsense. Luis Alvarez, perhaps the greatest experimental physicist of this century, remarked to me that the advocates of such schemes were 'very bright guys with no common sense.'"

Clarke continued: "Looking into my often cloudy crystal ball, I suspect that a total defense might indeed be possible in a century or so. But the technology involved would produce, as a by-product, weapons so terrible that no one would bother with anything as primitive as ballistic missiles."

In *Engines of Creation*, Eric Drexler proposed that we build an active nanotechnological shield – a form of immune system for the biosphere – to defend against dangerous replicators of all kinds that might escape from laboratories or otherwise be maliciously created. But the shield he proposed would itself be extremely dangerous – nothing could prevent it from developing autoimmune problems and attacking the biosphere itself.

Similar difficulties apply to the construction of shields against robotics and genetic engineering. These technologies are too powerful to be shielded against in the time frame of interest; even if it were possible to implement defensive shields, the side effects of their development would be at least as dangerous as the technologies we are trying to protect against.

These possibilities are all thus either undesirable or unachievable or both. The only realistic alternative I see is relinquishment: to limit development of the technologies that are too dangerous, by limiting our pursuit of certain kinds of knowledge.

The new Pandora's boxes of genetics, nanotechnology, and robotics are almost open, yet we seem hardly to have noticed. Ideas can't be put back in a box; unlike uranium or plutonium, they don't need to be mined and refined, and they can be freely copied. Once they are out, they are out. Churchill remarked, in a famous left-handed compliment, that the American people and their leaders "invariably do the right thing, after they have examined every other alternative." In this case, however, we must act more presciently, as to do the right thing only at last may be to lose the chance to do it at all.

I frankly believe that the situation in 1945 was simpler than the one we now face: The nuclear technologies were reasonably separable into commercial and military uses, and monitoring was aided by the nature of atomic tests and the ease with which radioactivity could be measured. Research on military applications could be performed at national laboratories such as Los Alamos, with the results kept secret as long as possible.

The GNR technologies do not divide clearly into commercial and military uses; given their potential in the market, it's hard to imagine pursuing them only in national laboratories. With their widespread commercial pursuit, enforcing relinquishment will require a verification regime similar to that for biological weapons, but on an unprecedented scale. This, inevitably, will raise tensions between our individual privacy and desire for proprietary information, and the need for verification to protect us all. We will undoubtedly encounter strong resistance to this loss of privacy and freedom of action.

It is now more than a year since my first encounter with Ray Kurzweil and John Searle. I see around me cause for hope in the voices for caution and relinquishment and in those people I have discovered who are as concerned as I am about our current predicament. I feel, too, a deepened sense of personal responsibility – not for the work I have already done, but for the work that I might yet do, at the confluence of the sciences.

But many other people who know about the dangers still seem strangely silent. When pressed, they trot out the "this is nothing new" riposte – as if awareness of what could happen is response enough. They tell me, There are universities filled with bioethicists who study this stuff all day long. They say, All this has been written about before, and by experts. They complain, Your worries and your arguments are already old hat.

I don't know where these people hide their fear. As an architect of complex systems I enter this arena as a generalist. But should this diminish my concerns? I am aware of how much has been written about, talked about, and lectured about so authoritatively. But does this mean it has reached people?

My immediate hope is to participate in a much larger discussion of the issues raised here, with people from many different backgrounds, in settings not predisposed to fear or favor technology for its own sake.

MONOGRAPHS

Asimov, Isaac: *I, Robot* (New York, 1950)

Garrett, Laurie: *The Coming Plague: Newly Emerging Diseases in a World Out of Balance* (New York, 1994)

Kurzweil, Ray: *The Age of Spritual Machines* (New York, 1999)

Leslie, John: *The End of the World: The Science and Ethics of Human Extinction*, (New York, 1996)

ARTICLES/ESSAYS

Clarke, Arthur C.: "Presidents, Experts, and Asteroids," in: *Science*, 5, June 1998. Reprinted: "Science and Society," in: *Greetings, Carbon-Based Bipeds! Collected Essays*, 1934–1998 (New York, 1999), p. 526

Forrest, David: "Regulating Nanotechnology Development" (Cambridge, 1989) http://www.research.lifeboat.com/forrest.htm

Joy, Bill: "The Future of Computation," in: *Nanotechnology: Research and Perspective*, eds. Crandall, B. C.; James Lewis (Cambridge, 1992)

Kauffman, Stuart: "Self-replication: Even Peptides Do It," in: *Nature*, 382, 8, August 1996, p. 496, www.santafe.edu/sfi/People/kauffman/sak-peptides.html.

FILM

Else, Jon: *The Day After Trinity: J. Robert Oppenheimer and The Atomic Bomb*, New York 1980, www.pyramiddirect.com

JAN VAN ROSSEM

THE BALANCING ACT OF DESIGN
ORIGINALITY, INDIVIDUALISM, MASS PRODUCTION

A crowd of incredulous visitors gathers in amazement around a simple wooden table in Ingo Maurer's Milan showroom. On it, an apparently equally simple construction: a thin metal rod, its upper end twisted into a thread holding a light bulb. The bottom end of the rod is clipped to a sheet of paper with a shiny metallic circle on it that is not completely closed. With a stroke of his pencil, the young designer connects the open ends of the circle and the bulb lights up. Magic? The designer smirks. A few particularly keen spectators examine the underside of the table for hidden switches. Nothing. So how does it work? The next and by no means less intriguing question: how do you switch it off again? With a rubber, of course, isn't it obvious? This apparent piece of sorcery goes by the name of "Watt". Simple trick, major impact. Young London designer Paul Cocksedge has simply taken advantage of the fact that even a tiny amount of graphite conducts electricity. And graphite, as everybody knows, is what pencil leads are made of.

A nice idea. But definitely not the kind of thing that will be gracing every second home any time soon. And yet Cocksedge's experiment says a great deal about the nature of design. What design wants, what it is capable of, what it ought to do. The design of this century, at any rate. For today design, what we call good design, has to accomplish very different things than what was expected of it in the past. For all its functionality it must still manage to surprise and entertain us, it must be inventive and may be ironic, it is supposed to evoke emotions and underline the user's personality. In the best possible scenario, it should in fact accomplish all of this at the same time.

Design doesn't have it easy nowadays: for all its functionality, it has to surprise and entertain us, it must be inventive and ironic and, as if that weren't enough, evoke emotions as well. Another goal is to increase the value of objects so that they can be sold at a higher price. The kitchen utensils by Alessi are a good example: although some are only moderately practical, what they all have in common is an unusual shape. Million sellers like the VW Beetle, Richard Sapper's "Tizio" table lamp or the Panton Chair are extremely rare. Most designer pieces only make it to the market in modest numbers. The expert opinion is that design should improve our quality of life – a goal that has long ceased to interest designers, according to Jan van Rossem. Apart from a few exceptions, virtually no designer since Verner Panton has really improved sitting comfort. Today a chair is supposed to evoke emotions and

DESIGN AS A MATRIX FOR MASS PRODUCTION

In the past the guidelines were simpler. At the outset of the previous century, when mass production was in its infancy, patterns were required for the unskilled workers on the assembly lines and machines, matrices that could be used for reproduction on a massive scale and without the need for too much expertise. The people that created these patterns were subsequently referred to as designers. A little later on, some designers made it their ambition to provide the people (i.e. the masses) with cultural commodities of high aesthetic appeal. The movement called itself Bauhaus. Yet even today, its products are enjoyed not by the masses but the well-to-do.

For decades on end, the goal of ambitious designers was to present an ingenious product that sold on a huge scale. Achille Castiglioni, the *grand seigneur* of Italian design, always described a nondescript switch as his favourite design. Not his fantastic lights such as the curved "Arco" lamp or the mischievously misappropriated tractor seat "Mezzadro". No, it was this switch that held the greatest appeal for him. Because not just one but several of these mundane objects made its way into virtually every household in Italy. A million seller.

The ideal design product is in fact expected to do the splits, to execute an impossible balancing act: innovative and avant-garde, it is nevertheless required to turn into a bestseller. A few excellent examples immediately spring to mind: the VW Beetle or Richard Sapper's halogen table lamp "Tizio", which was so avantgarde that it spent ten years languishing on the shelves until the time was finally ripe and it started selling in its millions. This list definitely also includes the Panton Chair, the last really new type of chair you can actually sit on comfortably. Although its popularity waxes and wanes, it sells in considerable numbers during its many boom periods (such as the current one).

look good, either standing alone or in a group. The approach taken by the latest generation of designers is both playful and serious. The idea is preceded by intense analysis of the problem. More than ever before, the author maintains, there is a demand for originality, as expressed not only in the form of semi-hand-crafted objects in small editions but also in the huge number of permutations for car equipment and features. Complex forms that can only be produced in small numbers and using special procedures provide protection from plagiarists. Contemporary design must be original and tell stories, insists van Rossem, citing a vase that changes into a lamp as soon as a single flower is placed in it. Just a gag? Perhaps. But also a good illustration of what design should do: analyse and find creative solutions.

In this context we are not talking about products like the iPod and its relatives. Let's face it: how much of the iPod's success can actually be attributed to the design? Do people really buy it on account of the charmingly rounded corners? And the ingenious design idea of a white casing with white earplugs only managed to sustain its cult status for two or three years. The iPod Nano has been available in black right from the start. The secret of the little MP3 player's success is not to be found in its shape or colour. Perhaps not even in its easy operation. Perhaps it is the inner values that really count: its capacity, the software technology from iTunes. And definitely the brand's cult status. Everybody wants to belong.

Whatever: with the iPod family, Apple launched a commercially successful mass-produced article. Most of the designs that experts can recall, those that remain in the collective memory, have only sold in modest numbers. Even Apple could be cited as a reference for this theory. However, the low sales figures probably result less from the avant-garde (and therefore not mass-compatible) design of the computers and more from the competitor's economic supremacy and superior marketing strategies.

WHAT REMAINS IN THE COLLECTIVE MEMORY?

And who can remember any particular pieces of furniture (except perhaps the Eames chairs)? Who can describe and possibly even name them? A few outstanding representatives of the Pop Art era, the classics by the avant-garde Italian firm Gufram, like the "Cactus" coat rack (which is useless for more than three jackets or coats) or the huge artificial meadow "Pratone" that invites you to flop down and relax. Some might even recall the ironic "Joe" armchair named after baseball star Joe di Maggio, a baseball glove blown up to armchair proportions. Others might even be able to name and recognise the "Swan", "Egg" and "Ant" designs by Denmark's brilliant Arne Jacobsen or, as already

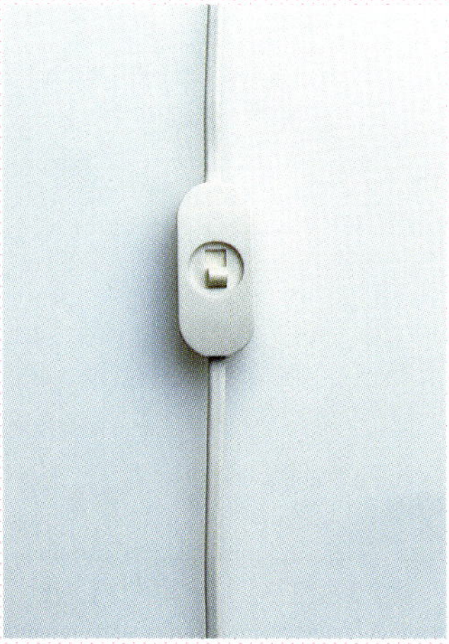

A million-selling was designer Achille Castiglioni's favourite design because it is used for its functionality, not for its designer name or because it feels good in the hand.

suggested, several pieces by Eames, the Panton Chair, Richard Sapper's "Tizio" and Philippe Starck's lemon squeezer – but that's about it. Except, perhaps, for the classics that return to the public consciousness as a result of one of the many "revivals". Yet even they are subject to fashionable fluctuations in public perception and regard.

So what is it that constitutes good design? If you ask an entrepreneur, a marketing manager, he will answer that its prime objective is to promote sales. And of course it should help boost profits. Its special exterior should justify its special (read: higher) price. Very often, this is precisely what happens. For years now, firms such as Italian kitchen utensil makers Alessi have been demonstrating how unusually shaped products can make huge profits. Fine, provided the consumer apparently doesn't care if the doll-like corkscrew "Anna G." by Alessandro Mendini is only moderately suited to opening wine bottles or if that most famous of all useless juicers, Philippe Starck's "Juicy Salif", squirts juice everywhere – a small amount of which even makes its way into the waiting glass.

DESIGN SHOULD ENHANCE QUALITY OF LIFE

Ask a consumer or a humanist and they'll advocate the most common and widely accepted expert opinion: design's task is to enhance quality of life. That applies to sofas, for instance, which are meanwhile obligingly available with deeper seats as a tribute to the chronic growth of humanity (in centimetres, that is). However, the seats are often lower as well, sadly ignoring the fact that the population is aging and finds it difficult to arise from such depths. Concerned about the back problems of today's far too sedentary work force, new office chairs are using flexible yet stable backrests to provide support whilst enabling freedom of movement. In this respect they improve our quality of life. But most of the products presented by designers fail to bring about any major improvements in comfort and convenience – most do not even try. Ever since Verner Panton's brilliant coup in 1960, designers (all designers except those that concentrate exclusively on lamps) have been trying their hand at chairs in vain – apart from a very few exceptions.

For a long time now, most designers haven't seemed that bothered about how comfortable the chairs are to sit on. For a chair is not merely a structure upon which to sit, usually at a table. A chair has to be capable of more than providing comfort and an agreeable opportunity to take the weight off your feet. It has to trigger emotions as well. After all, even a solitary specimen has a not inconsiderable effect on the appearance of a room. Surrounding a table in groups of four, six or even more, its impact is significant.

So what can be done, apart from making the sofa a little deeper, the drill a little more ergonomic, the door handle a little easier to open? Perhaps the latest generation of designers has come up with a good answer. They are taking a playful approach to their task without abandoning seriousness. For them too, the idea is preceded by intense analysis of the problem. An example: what is the most annoying thing about a vacuum cleaner? Answer: lots of things. True. The problem with the bags has been remedied by Mr. Dyson and a host of imitators. (Will this lead to the elimination of an entire branch of industry? The production of vacuum cleaner bags? But that's a different story.) Now there are even little vacuum cleaner robots that circle the apartment day after day but can't get into the corners. Whilst their superficial endeavours might hold a certain appeal for technology freaks, they certainly won't be winning any cleaner-of-the-year awards.

THAT'S ENTERTAINMENT!

At the end of the day, there's still no choice but to get on and do it yourself – and that isn't about to change. With or without a bag, vacuuming is a strenuous and boring job. Everybody knows that. And yet hordes of designers have failed to do anything about it. At least until Italy's Lorenzo Damiani finally took the necessary action and developed the "Airpouf", an appliance that tackles both problems with an eye to enhancing the user's quality of life. The entire body of the vacuum cleaner is upholstered (hence "Pouf"), transforming it into a pull-along seat. On top of that, it doesn't need to be stored in a cupboard. Having performed its duties, it doubles as a stool or footrest. When not in use, the hose and exhaust pipes are plugged with little balls of fabric in contrasting colours. A particularly nice detail: the exhaust pipe points upwards. When the vacuum cleaner is running, the little ball dances half a metre above it in the jet of escaping air and accompanies the appliance on its journey through the apartment as if attached to it by a magic ribbon. That's entertainment – while you're vacuuming. Hard to imagine, but true!

There is of course some justification for the objection that this is nothing more than a gag: at the end of the day, once you've vacuumed up for the third, fifth or tenth time, it's still only vacuuming and the dancing ball might no longer cheer you up as you perform this loathsome task. Maybe. But it's definitely worth a try. And don't forget the additional benefits: apart from the extra seating, just think of all the storage space it frees up. More than anything else, isn't design supposed to be useful?

Italian Lorenzo Damiani developed the first completely upholstered vacuum cleaner, the "Airpouf", which becomes a pull-along seat while vacuuming and a handsome piece of furniture when not in use. The accessories can be stored out of sight inside the appliance. Small fabric balls serve as plugs and dance in the jet of escaping air.

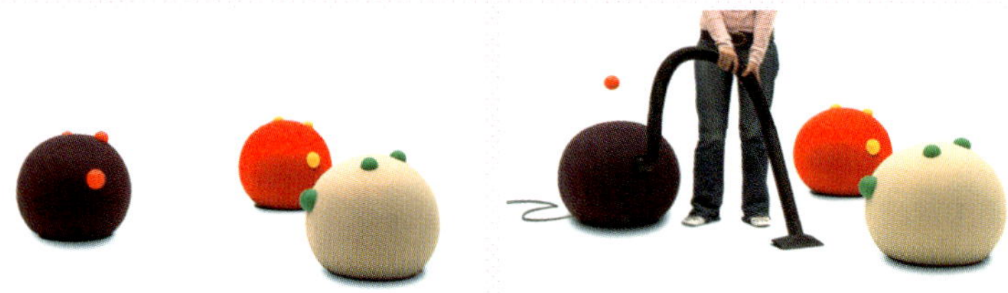

But perhaps there's another added value that design needs in our day and age. It ought to have a history, it should tell a story and establish a personal relationship with the potential owner. It should bewitch him with its charm and seem custom-made. That's what jeans are like these days. True, it might be a bit difficult to find the right ones, the right fit, amongst the vast array of models. But once you find them, you've got a pair that look as if you've always had them. Even though there are hundreds of identical specimens in every size. But not countless thousands. The very nearly individual markings left by the washing process establish a personal relationship to the wearer.

THE INDIVIDUALISATION OF MASS PRODUCTION

More than ever before, design is about originality. As contradictory as it might sound, it is about originality in the sense of the new, the unique. On the whole, it is about the exact opposite of production on a massive scale. Italian doyen Gaetano Pesce is one of the pioneers of this theory. His products, such as the "Nobody's perfect" series of furniture, don't even want to be considered beautiful. First and foremost, they are completed experiments on the individualisation of mass production. (Although when it comes to this series, "mass" is hardly the right word.) No two chairs are alike, it is a series of models made up of a host of one-offs. Pesce even encourages workers at the furniture factory to bring their own creativity to the production process and change colours, patterns and details of the shape.

Which puts him right in touch with the times. Take today's cars, for instance: it would be surprising to find two totally identical vehicles. BMW has calculated that, for several of its models, the various engine sizes, technical specifications, equipment features, colours, fabrics and interior designs could theoretically result in more than a million variants – far in excess of the number of 5 Series cars actually built.

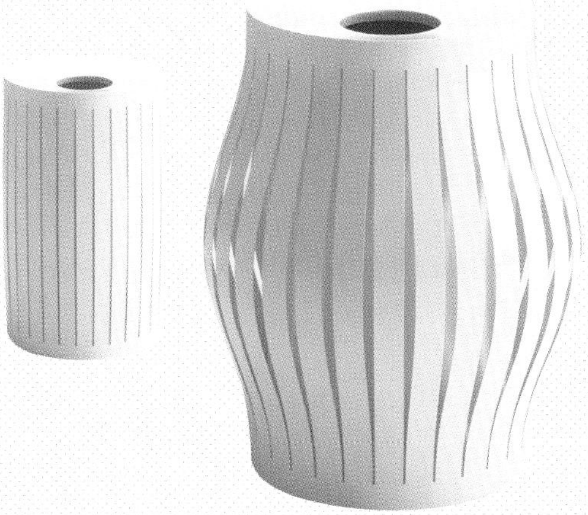

The Swedish design group Front brought the principle of the digital computer screen trashcan back to the analogue world. When your trashcan basket is full it bulges outward. When half-full, it is in permanent motion. The internal basket pushes downward while the external slats try to maintain an upright position.

There is another aspect, too: it isn't just that not even an expert can unequivocally identify a Vitra chair as an original any more, despite the stamp and certificate – all that is easy enough to copy these days. For some time now, any Chinese or Korean manufacturer has been able to copy any sofa, any chair he likes, whenever he likes and in any size, shape or colour he likes. Today preserving originality means coming up with a design that is not geared to mass production and mass tastes at all but which is instead so complicated and special that it can and is only intended to be produced in small numbers – making it uninteresting to copyists of any kind.

Two pairs of brothers serve as excellent examples of the different approaches constituting design. Firstly Humberto and Fernando Campana from Brazil, who use recycled scraps to create bizarre furniture in exclusive editions. Their designs are in fact more like artistic sculptures. Take the "Favela" chair, for which different-sized pieces of wood are nailed together by hand to resemble an improvised slum building. French brothers Ronan and Erwan Bouroulec, who have devoted themselves to classic industrial design, represent the other extreme. Their speciality: tiny plastic units. In sufficient quantities, they can be connected to create walls and even miniature architecture. With this kind of thing, however, the copy is often faster than the original. Even before more than a select few were aware of the "Algae" produced by Vitra, other firms had cheekily presented their own – let's call it "interpretation" – of the room-divider elements.

The most recent design by one of the latest Swedish designer groups is unlikely to fall foul to the same fate. The four young ladies, all in their twenties, go by the name of Front. But not in the war-like sense of the word: in Swedish, front means at the fore, avant-garde. At the Milan Furniture Fair in April 2006, they presented a bin whose appearance and function are reminiscent of the trashcans familiar from computer desktops. The white metal cover sports lateral slits every few centimetres. The trash container hangs

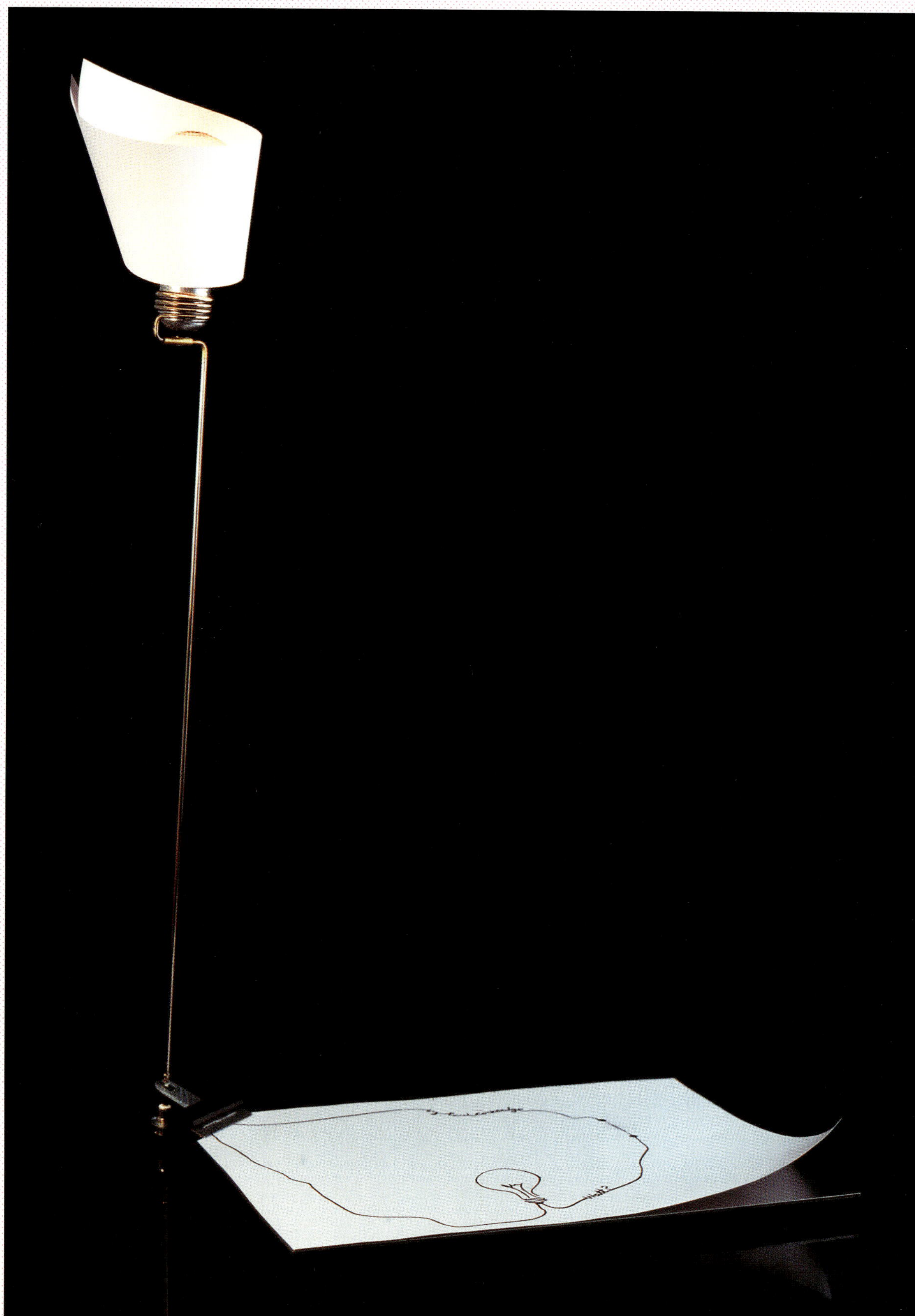

Young London designer Paul Cocksedge's lamps are surprising in their playful finesse. With "Watt", a final pencil line drawn on a piece of paper completes the electric circuit and makes the bulb light up.

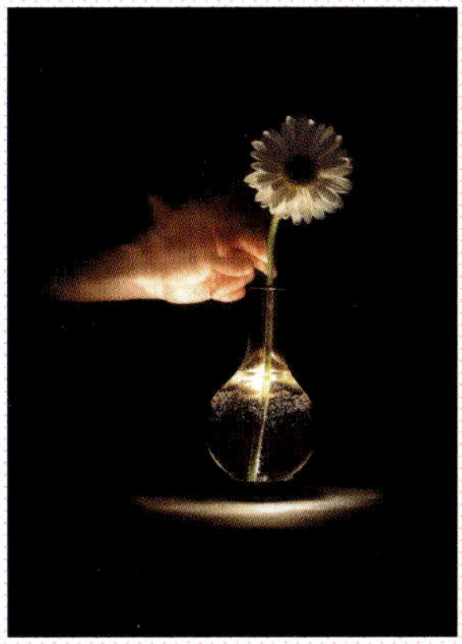

Nor will you find a switch à la Castiglioni on Cocksedge's "Bulb" lamp. Concealed wires on and in the glass conduct electricity. As soon as a single flower is placed inside the vase and touches the upper metal edge, the circuit is completed, lighting the flower from the base.

inside and is pushed downwards under the weight of what is thrown away. When full, the bin becomes distended – just like on the monitor. In a long hotel corridor, this means staff can see whether the bin needs emptying even from a distance. The trashcan – surprise, surprise – even has a certain entertainment potential, in line with the modified maxim "Form follows FUNction". When half-full, it remains in permanent motion: the rubbish container inside pushes downwards, the external slats try to take up their elongated basic position. This ingenious interplay keeps the outer casing in motion: it seems to be taking a succession of deep breaths.

ORIGINALITY NEEDS A STORY

In this way a trashcan is given human traits, it becomes a talking point and tells a story – in this case a story developed from its function. More than ever before, contemporary design is linked to its context, as the products by Paul Cocksedge, the designer mentioned at the start of this essay, illustrate. He sold hundreds of his "Bulb" lamps to a restaurant on the Canary Islands, where simple glass vases adorn the tables. When the waiter brings the menu, he doesn't light a candle but instead removes the flower from his jacket pocket and puts it in the vase, which immediately lights up. Concealed wires on and in the glass form a not quite closed circuit, which is only completed when the single flower is placed in the water and leaned against the metal upper edge of the vase. It conducts the electricity and makes the bulb in the base of the vase light up.

Another example: the "H2O Chair" by Barcelonan Martí Guixé, who refers to himself as an "ex-designer". A chair with a very high back rest and very, very short legs. Guixé designed it for the owner of the H2O Gallery in Barcelona, who had established a reputation for himself as a bookworm. Which is why the sitting height of the chair can be varied by stacking books on the seat. The operating instructions are immortalised on

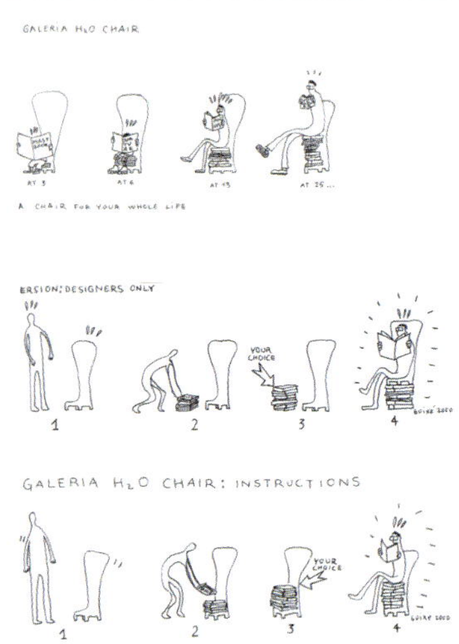

The "H2O" chair by Martí Guixé of Barcelona has a very high back and very short legs. The sitting height can be adjusted by varying the number of books and magazines.

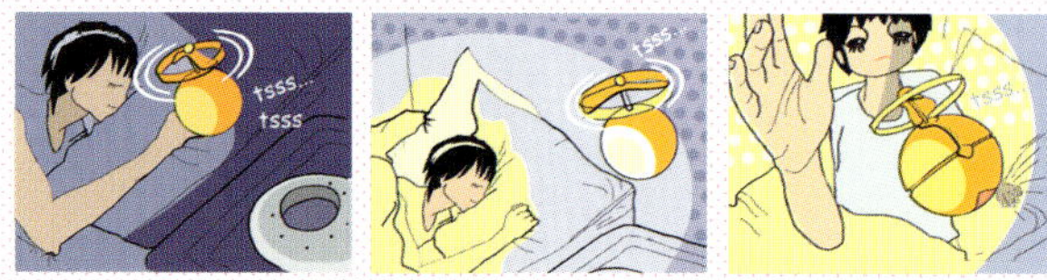

Ena Macana and Nadine Meisel's "Blowfly" alarm clock does not ring. It buzzes around the sleeper like an annoying mosquito. You have to get up and catch it to shut it off. A manufacturer has yet to be found.

the side of the chair in the ex-designer's typical humorous line drawings. An expert version is also available, requiring the user to put the chair on top of the books.

DESIGNERS HAVE TO BE INVENTORS

More than anything else, this is of course a gag. But it also illustrates the main tasks of design: analyse, consider contexts, find creative solutions to individual needs. Which is precisely what young designer Nadine Meisel did: her latest design revolutionises the detestable necessity of getting up in the mornings. All alarm clocks have a common weakness: they are too easy to turn off, considerably increasing the likelihood of that blessed state of oversleeping. Nadine Meisel identified the problem and drew the following observations: firstly I need something that really wakes the sleeper up, and secondly I have to make it as difficult as possible for him to turn it off. Instead of ringing, her "Blowfly" alarm clock takes off on a circular flight above the sleeper's head, buzzing like an irritating mosquito. That's what wakes you up. You can only turn it off by getting up and catching it. Which takes you out of the horizontal plane and ensures you stay awake. The alarm clock uses the latest technology from Epson, whose developers no doubt had rather different purposes in mind. But it might well become a big seller. After the initial publications, the designer was inundated with hundreds of orders. A Harry Potter fan club even suggested using the model in the next film about the apprentice magician. Which leaves just one small problem: the product has yet to awaken the interest of a producer.

www.campeggisrl.it/damiani.html --- www.frontdesign.se --- www.guixe.com
www.paulcocksedge.co.uk --- www.nadinemeisel.de --- www.blowflyalarmclock.net

WOLFGANG PAUSER

-------THE PROMISE OF HAPPINESS: "EMOTIONAL DESIGN"-------

"Except ye be converted, and become as little children,
ye shall not enter the kingdom of heaven."
(Mathew, 18:3)

It is time to announce the birth of a new science even younger than the recently inaugurated millennium, a science which, if its name is to be believed, takes an exceedingly positive view of the future: "Funology".[1] The Anglo-American name already hints that the cultural environment to which it owes its conception has no intention of engaging in scrupulous feuilleton debates on how to evaluate a "fun society" but is instead determined to grab happiness vigorously by the pigtails, even if that means subjecting both the pigtails and the head to which they are attached to a thorough trim and sprucing-up first. Yes, it's design we're talking about – in particular, the theory design of the latest design theory: "Emotional Design".[2]

The combatants of the "joy-of-use" debate have already organised, there's a "Design and Emotion Society"[3] and international congresses[4] that aim not only to "turn fun into serious stuff"[5] but to give concrete shape to the foundation and structure of the new science that will finally guide us device-embedded humans to happiness by the light of methodical design. Terms like "hedonistic quality", "emotional usability", "pleasurability", "aesthetically pleasing", "mood", "play" and "fun" are still competing for the status of fundamental concept. So what if they all mean the same or articulate tricky differences in the way we enjoy objects – the goal of all this epistemic endeavour is as obvious as the

Marketing experts are currently stylising "emotional design" as the ultimate recipe for success. Forms that give joy are allegedly better to use and preferred by consumers. If it were possible to objectively ascertain which forms trigger positive emotions and thus the impulse to purchase, wouldn't all design problems be solved for ever? Wolfgang Pauser believes this approach is far too simplified. Merely portraying a smile or appealing to forms with a positive effect on our cerebellum is not enough to trigger the reaction of a smile. Emotionalised communication is only capable of boosting turnover for a few products, markets and sectors. As a patent remedy, "emotional design" fails because we are not only controlled by the cerebellum but by higher, reflective cerebral structures as well. Perhaps, following the example set by advertising, emotional elements in design should remain more subliminal in

criterion of its empirical measurability: a smile on the user's face as a response pattern produced by the form. No matter that the new "Mini Cooper S", joy-of-use researchers' favourite design example, has a host of functional deficits: no other car has conjured up a smile on the faces of so many people, argues "design guru" Donald Norman in his book *Emotional Design*.[6] Like the bible, this particular publication is currently at the centre of a debate, nay, hype, a fashionable madness, it is the focal point of a design ideology that is proving highly successful in current industrial design practice.

Yet who or what exactly is the bestselling American author arguing against? Mainly against himself, against the author of *The Psychology of Everyday Things* from 1988. Back then he made a strict case for functionality and reproached many designs for their inadequate practicability. "This was Nasty Norman, now he's become Nice Norman, who smiles and tells you how great everything is," remarked *The Guardian*[7] on the publication of *Emotional Design*. "This is my new life" the author himself says in explanation of his personal transformation from Saul to Paul, from rigid functionalist to fun preacher. Two empirical facts led to this conversion: firstly he had to admit to himself – evidently an embarrassing confession – that, in the privacy of his own home, he collected things he did not buy for the sake of their function but out of pure joy and love. Secondly, he was only able to resolve the contradiction between his professional role as professor of behavioural psychology and stern propagandist for functionalism and his secret private passion when a scientific discovery proved what had hitherto been not only impossible but impermissible: it was empirically proven that people can handle the function of objects better, faster, more productively and more intelligently when these objects are enjoyably, attractively, pleasantly and pleasingly designed, i.e. when their form results in joy. Joy is an emotion! And if it contributes to better functioning, it should no longer be denied the right to exist. Let that suffice on the myth surrounding the birth of the new science of "funology".

order to seduce, suggests Pauser. Or is marketing's dream of designing emotions by means of certain forms nothing but wishful thinking? Because cultural, symbolic and social differentiation systems are also involved in our perception and evaluation of forms, we do not react in the same predictable way as laboratory rats. The term "emotional design" is often abused in advertising copy for cars. Pauser concludes that, amongst other things, it serves to make design's achievements plausible to form-blind managers and, for instance, allocate a vehicle to a higher price segment. The term seems more appropriate in relation to products such as shavers. But it remains doubtful whether design's task is to illustrate the structures of our unconscious in the form of things.

It is certainly nice when a behaviourist can smile for a change – and in doing so can embrace the emotion of joy all the more because the narrow world of his own empiricism has provided a good reason to legitimise it. But Don Norman is not content merely to celebrate his new-found ability to abandon his resistance to emotion (an occupational hazard in his profession). No, he resolves to draft a theory of the emotional effects of design. Since it was beautiful things that led him to discover his access to feelings, it is certain that emotionally moving design already exists. The spider-legged lemon squeezer by Philippe Starck adorns the cover of his book – its purpose not to extract juice but to initiate conversation, now a thoroughly legitimate function. The legendary E-Type Jaguar of 1961 still evokes emotions today. People prefer to put flowers in bulbous mineral water bottles rather than straight ones. What enables a particular form to trigger emotions in us, independently of any function it might serve? If an empirical science could objectively ascertain and prove which forms give rise to positive emotions, wouldn't all design problems be solved for ever? Unlimited joy-of-use for consumers, unlimited use-of-joy for industry. What brilliant interaction becomes feasible, for the common good, nay, for the greatest happiness for the greatest number!

Before the cheering gets out of hand, let's return to the theory. Simplicity is an essential aspect of American theory and bestseller design. How many elements does the design emotion have? That's right, three! Because, of course, the brain also consists of three interdependent parts. Which means we can ascend from the lowest to the highest. According to Norman, the information received by the sense organs is already given a positive or negative emotional "hue" by the vegetative nervous system before it reaches the more complex instances of judgement in the higher hemispheres of the brain. And recognition is inseparable from this "visceral", subconscious feeling. What humans have in common with earthworms is, for Norman, the actual seat of what he calls emotion and

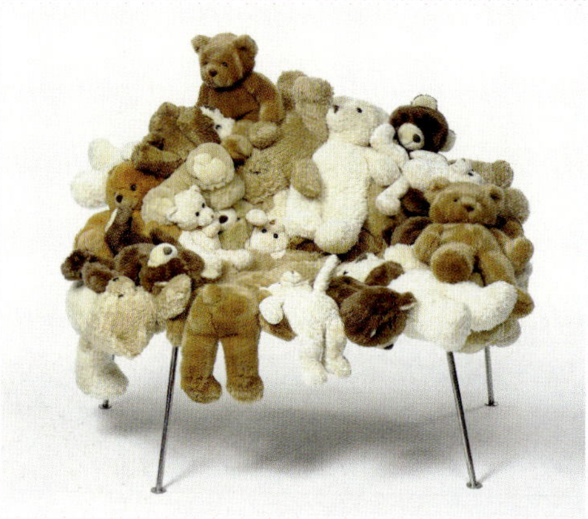

A response by design to Donald Winnicott's theory of the transition object? "Teddy Bear Chair" created by Brazilian designers Fernando and Humberto Campana (2004).

what he proposes to research as a viable source for the derivation of future design principles. So far this approach has borne decidedly scant and not particularly fresh fruit. When the Smart's radiator grille smiles, it evokes the emotion of cheerfulness in the onlooker. In car design, it isn't only frontal similarities with a face that play a role: there are metaphors involving other body parts too. Colours can influence moods, smells stir up memories. The baby schema and other key stimuli pull at our heartstrings, and isn't round generally more positively emotional, whereas anything pointed and angular has a detrimental effect on the "joy-of-use"? Babies don't like grey right angles, therefore...

Joy occurs when a human and his relationship to form prove as simple (to control) as a laboratory rat. Or that, at least, is when marketing managers experience joy. For this is strictly scientific proof of both the necessity and reason for cars to smile or snarl, for machines to be decorated with mock muscles or exhibit feminine curves, for cute little appliances to appeal to our maternal instincts or be as garish, round, eye-catching, animal-like and animistically animated as possible. But beware: isn't disgust a human emotion too?

"Insofar as they embody the identification of people with their own debasement, the grinning cultural commodities are vulgar... One of the defining elements of kitsch may well be the simulation of nonexisting feelings and thus their neutralisation along with the neutralisation of the aesthetic phenomenon."[8] It would be easy to dismiss Adorno's disgust for grinning design as the cultivated posturing of a past era, were it not for the fact that marketing experts themselves are concerned about how all too direct access to the cerebellum functions. Merely portraying a smile – as in classic advertising that uses models as mediators to emotionalise the product – is not in fact enough to trigger the reaction of a smile. Even if it's true that laughter is often infectious and appealing, there is no

An airline as ambassador for an all-embracing emotional design strategy: All Nippon Airways painted its aircraft with "Pokémon" monsters and sold merchandising articles on board.

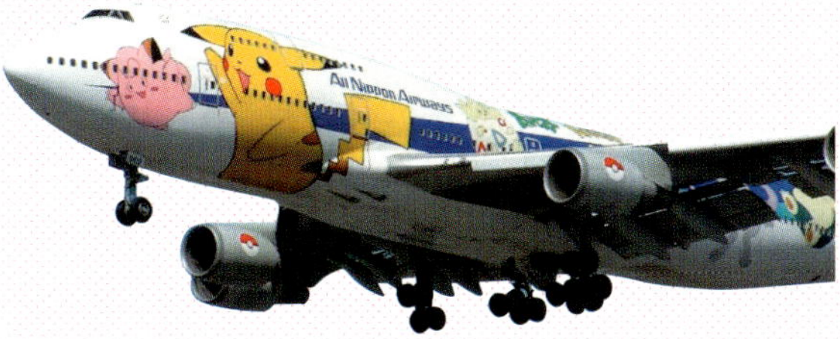

guarantee that design can transport emotions as if by some law of nature. The sales-promoting effect of emotionalisation in marketing, the current credo and management's last great hope, is relativised by a new McKinsey study of 80 brands[9]: emotionalised communication is only capable of boosting turnover for certain products, markets and sectors, for others it has long since proved counterproductive. Perhaps the cerebellum is the only instance that takes pleasure in colourful, grinning, bulbous products, then to be left pretty much alone with its joy. Disgust, certainly a "visceral" emotion, is not only triggered by food gone bad, a phenomenon with a perfectly simple phylogenetic explanation; as psychoanalysis has shown, it also occurs as a form of defence against objects we used to be particularly close to but have now freed and distanced ourselves from as we mature. The planability of emotional reactions to design that appeals to the cerebellum is severely impaired when not only the visceral nervous system is involved in the recognition process but the higher hemispheres as well, when all their complex cultural and memory functions are interwoven with nausea, so that a strict before-and-after effect in the perception process does not apply to those consumers who, in a matter of seconds, cannot help but become involved with the whole of their feeling-thinking-acting being.

Perhaps the "hidden persuaders", having been transferred from advertising to product design, have to remain secret and subliminal in order to seduce. Then commercially orchestrated access to the visceral dimension of the subconscious would merely make its blatant factors visible and conscious but, by transferring them into formal brazenness, would rob them of their impact or, to put it another way, eviscerate them. Once the marketing managers get hold of the figures, that certainly wouldn't be the worst thing that could happen. More than two hundred years ago, Immanuel Kant had already observed that the sound of a nightingale chirping at dusk held an aesthetic appeal for guests in the garden of an inn, an appeal which however immediately disappeared once they realised the innkeeper had hired a lad to imitate the bird and hidden him behind a bush. The consumer sees what's behind the bush and is annoyed. Could it be that, although design has always triggered emotions, marketing's dream of being able to design emotions with forms is based on a fundamental misunderstanding? Then the theory of "emotional design" itself would be nothing but emotionally designed sales rhetoric that aims to sell the seemingly empirical, scientific-sounding terminology to wishful-thinking managers whose greatest dream is for consumers to be as easy to understand as a laboratory rat. But when a company decides it wants to make its customers smile, the brand, marketing and design are subject to a very human law: a false smile, put on to please, will never have the same contagious effect as a genuine smile that bubbles up from the inside.

Norman of course also knows that there is more in our brains than vegetative, "visceral" reactions to stimuli – that there is more than what has been determined by natural history, which for the author also includes colourful plastic iMacs, blue skies and apple cake. For one storey higher up in the cerebral architecture lives "behavioural design", responsible for all aspects of smooth functioning. On the third floor of the theory architecture, close to the top of the skull, things start getting complicated: this is where "reflective design" arrives – and whilst the author only understands reflection as the image of the anxious-to-impress consumer in the eyes of his social environment, its interpreters occasionally seize on its ambiguity and include reflection in its extra-behavioural sense as an entry under design perception. At this higher level nature remains silent, "acquired tastes" must therefore be taken into account as well. They, says Norman, are responsible for the fact that people do not only eat sweets (something he acknowledges with amazement) but, contrary to their infantile nature, also prefer more differentiated tastes, even – can you believe it – putting salt and pepper on their food.

At this point it is pleasing to note that, besides the pure joy of the limbic system, cultural, symbolic and social differentiation systems are also involved in the perception and evaluation of form. However, the laboratory rat method cannot – and must not – be applied to the complexity that surfaces here. And so it is of little use when Norman concedes that, in addition to attractiveness, there is beauty as well, which exists "below the surface" as a result of conscious reflection and experience. For as soon as the design consumer is deigned a complex and reflective social being, all the models and terminology fail even to describe him. Whereas to understand laboratory rats all that's needed, if anything, is a better-designed experiment.

Time to move on from the recently arrived theory to the long-established practice of emotional design. In practice, it is hard to think of a contemporary car design that is not described as "emotional" in the advertising copy.[10] Since when conducting an experiment on myself I could ascertain no surge of emotion whatsoever at any level of my brain when confronted with the new Mercedes Transporter, I can only conclude that, while a new car model is in production, this sort of declaration in advertising copy has the function of

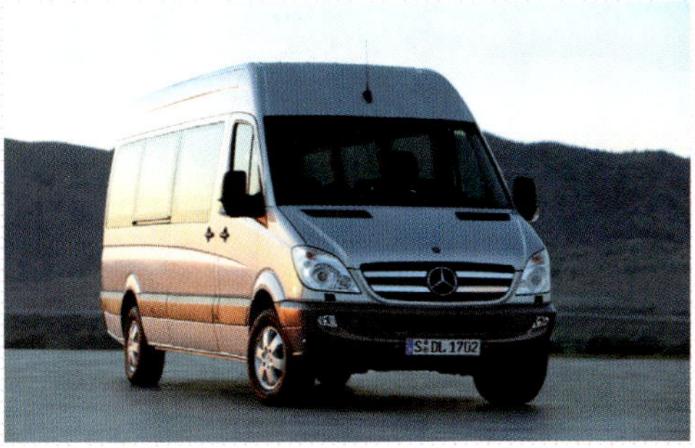

Thanks to Internet mail order sales, vans such as the Mercedes "Sprinter" are now seen everywhere, whether obstructing traffic or speeding in the fast lane. According to the advertising slogan, the "emotional design" of the new-generation vans is a real eye-catcher.

making partial design and textual performances plausible to form-blind managers who then consent to pay a higher price. Since I don't know anybody who reads advertising copy of his own free will, this particular field of success for the term "emotional design" can, with no further ado, be left to fizzle out in the non-too-distant future. After all, isn't the best thing about the hype surrounding fashionable formulas for happiness its rapidly approaching expiry date?

In practice, numerous manufacturers of industrial products have long been sneaking emotionalisation calculations and key stimuli into the design of their goods. The Gillette wet shaver for men, for example, goes by the name of "Mach 3", metaphoricising fast projectiles, aircraft and inherent potential (the "M3-Power" even comes with batteries and a motor!). Its counterpart for the opposite sex speaks of the goddess and has a handle shaped like the curved forms of the ideal female figure ("Venus Devine", in the motorised version the aircraft turns into a vibrator and is called "Venus Vibrance"). Yet is it possible to derive such gender imagery from Norman's theory? A smattering of Lacan helps: the man is missing a phallus, the woman is the phallus... The need for social and cultural background knowledge becomes even more acute when it comes to understanding the latest transition from three to four and finally five blades in the men's version. The current men's fashion for shaving their wrinkled scrotums in emulation of the porno stars and their visibility-enhancing measures that have made their way onto the bedroom telly makes a multi-blade system a very real necessity – for purely functional reasons. And yet this fashion for feminising the male body would never have asserted itself if shaving the male chest had not already been well rehearsed as part of the masculine-coded bodybuilding mania of the 1980s. Ever since, the design of the male chest – proudly protruding and hairless – has increasingly come to resemble that of the female chest, which can only be interpreted in the context of shifting social roles. The potency metaphors expressed

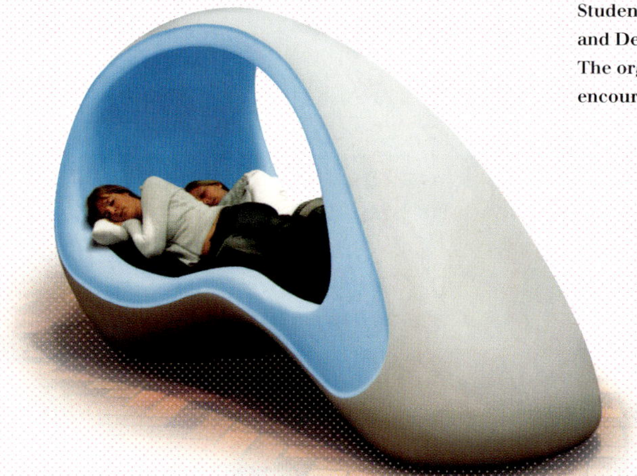

Students at Stuttgart University's Institute for Living and Design created the "Napshell" for restful napping. The organic shapes of the open sleeping capsules encourage users to cuddle.

in the design upgrade of the new shaver for the male body might have the emotional function of symbolic compensation for the imminent feminisation of the social male role. As for the impact this entire drama has on man's emotional relationship to his razor, it can only be described as ambivalent, situational, multi-dimensional and overly complex – and, if at all, in detail. Making implicit desires explicit can easily produce the opposite effect, as any man who has ever toyed with the idea of buying a conventional potency symbol like a red Porsche knows.

He is haunted by the fear that his excessive protestations might be interpreted as a question mark. When a woman displays a dozen cuddly animals on the rear shelf of her "Mazda Baby", this *in rem* divulgence of her secret interest in reproduction is equally embarrassing. It remains doubtful whether it is design's destiny to illustrate psychological theorems about the structures of our subconscious in the form of things. Following on from Marx, Freud and Lacan, philosopher Rudolf Heinz refers to the thing as an "unawareness lock" that characterises it in the psychoanalytical sense both in terms of its genesis in technical and economic history and as an object. This type of lock on selfhood is not easy to open, at most narrations and images can be applied externally in the attempt to make a supposedly unconscious inner life public. Yet what psycho-designed surfaces make legible is not the unconscious but the current model with which we interpret the emotional apparatus. Thus it is not in the history of science that Donald Norman will find his future place, but in the history of design. For no matter how illusionary a theory is, as a style-shaping misunderstanding it can become relevant to art history, when good designers believe in it or functionalise it as a legitimation discourse. Peter Behrens wasn't a bad architect just because he was inspired by crystal esotericism. And the Prague cubists left us with some astonishing works, even though their theory-dominated will to shape a new humanity by means of acute-angled forms sadly failed to have the desired effect. The right angle and the straight line were once supposed to make us happy as well – a clear case of emotional control by means of design! Unfortunately, this pathos led to the crude reaction of architecture à la Friedensreich Hundertwasser – whereas these days nobody believes that a crooked line is per se divine and makes us happy. Rounded shapes, on the other hand, whether they broadcast viscerally stimulating appeals or not, have always meant something totally different, whether to Boullée or Colani, in the 1950s or today's blob architecture.

What links design with emotions is its meaning. Forms do not have a meaning in themselves, but only as a result of their relationship to other forms. That is why context is the only thing that can give meaning to a thing's form and thus create an emotional relationship between a human being and the thing. That leaves no scope for positivism – a rather positive outcome.

1 Blythe, M. A., Overbeeke, K., Monk, A. F., & Wright, P. C.: *Funology: from usability to enjoyment* (Boston, 2003)

2 www.design-emotion.com; www.designingemotion.nl

3 Design and Emotion Society: www.designandemotion.org

4 "Design and Emotion 2006" conference in Göteborg: www.de2006.chalmers.se

5 Ahmet Cakir, quoting Inga E. Reeps, Alan Ross: *State-of-the-Art Analyse Usability, Design and Joy of Use* (Constance, 2004)

6 Norman, Donald A.: *Emotional Design: why we love or hate everyday things* New York 2004. www.jnd.org

7 http://technology.guardian.co.uk/online/story/0,3605,1166467,00.html

8 Adorno, Theodor W.: *Aesthetic Theory*, p. 397, translated by Robert Hullot-Kentor, quoted by Liessmann, Konrad Paul: *Kitsch! oder Warum der schlechte Geschmack der eigentlich gute ist.* (Vienna, 2002), p. 60

9 "Im Wechselbad der Gefühle", by Prof. Dr. Manfred Kirchgeorg, HHL-Leipzig Graduate School of Management, Tjark Freundt, Dr. Jesko Perrey, McKinsey & Company, published in *ABSATZWIRTSCHAFT* 06/2005

10 Mercedes-Benz Sprinter, Subaru, Honda Civic, Volvo YCC, smart, Audi, Canyon Bicycles, Glunz hardboards and many others

IN HIS "CRITIQUE OF JUDGEMENT", IMMANUEL KANT WROTE: "BEAUTY IS THE FORM OF FINALITY IN AN OBJECT, SO FAR AS PERCEIVED IN IT APART FROM THE REPRESENTATION OF AN END." YET FROM THE VERY BEGINNING, DESIGN HAS SERVED ENDS. THE ECONOMIC SUCCESS OF A CREATIVE IDEA IS AN END BOTH DESIGNERS AND THEIR CLIENTS STRIVE FOR. BUT DOES THAT MEAN DESIGN CANNOT CREATE ANYTHING BEAUTIFUL? – OR MUST IT, SHOULD IT STRIVE FOR BEAUTY AS WELL, IGNORING ITS ENDS AT LEAST OCCASIONALLY? THE RELATIONSHIP BETWEEN BEAUTY AND ECONOMICS IS CONTRADICTORY AND REQUIRES CAREFUL REFLECTION.

——— BEAUTY AND ECONOMICS ———

ROGER M. BUERGEL

-------GIVE BEAUTY A CHANCE-------

Beauty is not something immediate. A parable by Abu Hamid al-Ghasal (1059–1111) describes a painting competition between the Byzantine and Chinese peoples. The two groups of painters were assigned to work on opposite walls of a large hall. To prevent them from cheating, the hall was divided in two by a large curtain. The Byzantines brought magnificent pigments with them and began to paint. The Chinese did nothing more than polish their wall to perfection. At a certain point the Byzantines finished their picture. When the Chinese claimed to be finished as well, there was general astonishment. How could they have painted their picture without any paints? But the Chinese said, "Just raise the curtain and see for yourselves!" There on their wall, all the splendor of Byzantine painting was mirrored, only with greater brilliance and depth.

Should it surprise us that in the time of Hartz IV[1] beauty is on the agenda again? And if it does, then how should we respond? Should we interpret the yearning for beauty as a symptom of decline? As yet another lifeboat with which the middle classes – whom the philosopher Giorgio Agamben has already begun to apostrophize as the global petite bourgeoisie – are seeking to make their social and political downward slide a little bit more bearable?

There is no doubt that beauty – at least true [*wahre*] beauty (as opposed to the commodity [*Ware*] beauty) – should be defended against any critical gesture that regards the yearning for beauty as purely regressive and seeks to unmask it. For properly understood, beauty is not something to which one could possibly cling as a vehicle for one's own contemptible hopes. It is not that beauty is something completely intangible – that is certainly not the case – but simply that it does not reside in things. On the contrary, it is the

Every involvement with beauty presupposes an active involvement with the things that one finds beautiful. That means psychological work, for beauty is not something that can simply be consumed. The bourgeois middle classes have rediscovered the subject of beauty in recent years. In the era of Hartz IV, they are involving themselves with beauty in an effort to distract themselves from the threat to their own existence and the hopelessness that goes along with it, and attempting to make both of these things bearable. Beauty only has a chance when it can be joined with an idea. Beauty as such is difficult to thematize, especially for art. A flowery meadow is already beautiful – what need is there for artists to represent it? In Marcel Proust's book *Swann's Way*, there is a passage in which the young

result of an energetic investment. Beauty, in other words, is work, psychological work – it is a projection. That much is clear from the parable of the Chinese mirror image. Nevertheless, it is probably true that today, there is something like a prevailing political mood that harmonizes splendidly with the veneration of static concepts. And it would seem to go without saying that beauty is one of those concepts, as a timeless and probably also transcultural construct, in short, a fixation. Even the platitude that beauty is ephemeral – that it is eminently mortal – does not argue against this view, since death, as the actual aim in life of the global petite bourgeoisie, is the consummate standstill.

Perhaps the first thing to be said in beauty's defense is that it is more or less the opposite of meaning. Beauty, in a progressive sense – a sense that is opposed, for example, to the imperative of total communication – has nothing to say. And that means in turn that it cannot be made an object of knowledge, although it is possible to learn from it. This is not to ignore the importance of those philosophical traditions that have devoted so much effort and energy to beauty, above all early romanticism, which traveled the treacherous path between philosophical cognition and poetic revelation. It is simply to say that beauty is not an appropriate subject for thematic art exhibitions. Or at least not for art exhibitions that do not allow art to radiate, but tackle it with a streamlined academicism. *Über Schönheit (About Beauty)* is the title of an exhibition currently at the Haus der Kulturen der Welt (House of World Cultures) in Berlin, an exhibition about which one is free to think what one likes. For myself, I find the direct and all too conventional association of beauty with the naked female form every bit as tiresome as the affirmation of Shirin Neshat's Leni Riefenstahl aesthetic, which has been polished to a high gloss by the art market.

The real objection, however, has to do with the idea that beauty can be illustrated by reference to things. Things that – whether artworks or not – have some kind of con-

Marcel lets a meadow full of buttercups work its influence on him. More than that, he corresponds with the flowers, which mirror in him a lost, paradisiacal state. That is the kind of intense involvement with beauty that Buergel imagines.

nection with the concept "beauty". In Proust's book *Swann's Way*, there is a passage in which the young Marcel lets a whole army of buttercups work its influence on him. He lets them do so until he begins to radiate inwardly himself. This internal radiance is brought about by a peculiar coalition between Marcel's psyche and the appearance of the flowers. Marcel corresponds to the flowers, which mirror something back to him that he lost long ago, a paradisiacal state that he carries around with him, as all human beings do. Lost – that sounds tragic or at the very least nostalgic, but it is not meant that way. It is simply gone, sometime in very early life, this oneness with things. And it never comes back, except perhaps during good sex – or at a good exhibition.

However, it is only partially true that beauty is the opposite of meaning. It is also the basis of meaning. But to say so is already to evoke a different kind of meaning, one that is substantially more emphatic, more intense, but also easily more unbearable than that of a traffic sign. Within every meaning, the act of its production continues to resonate. That too is something that humanity learned at a certain point, that words and concepts do not refer to things themselves but merely exist in opposition to other words and concepts. The word "apple" does not mean apple, but simply "not a pear," "not a banana," and so forth. This is a commonplace, but at any rate it is a commonplace to which we owe an entire creative tradition. But I am digressing (unbeautifully). Beauty, to put it another way, is also the result of an affective assertion with which we determine things in the world. By finding them beautiful, we bring them out of their mere existence and make them meaningful.

A flowery meadow is marvelous all by itself. What do we need artists for?

To me, this capacity to bestow beauty seems to be an absolutely decisive means for opposing the general battering of our planet and its existence. Yes, flowers can be

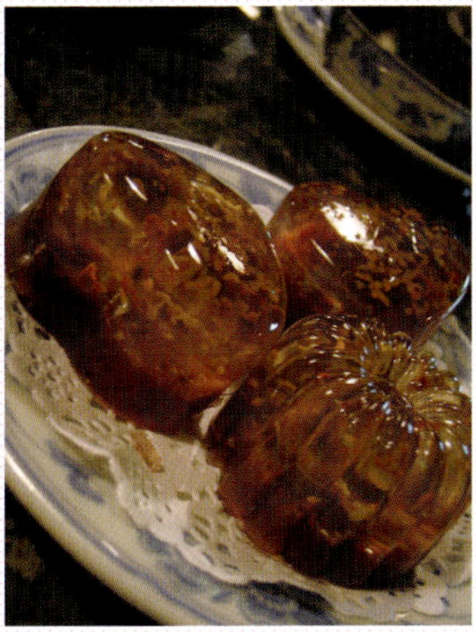

Michel Cluizel:
Dessert in Guangzhou

patented. Alternatively, one can respond to their luster and immerse oneself in the sight of them. Perhaps one day we will understand capitalism as a psychological defect, as the impotent and ultimately self-destructive attempt to feed on everything, in order to eliminate this damned lack, which is the basis of our existence.

Beauty, for its part, shows us how to live with this lack, or perhaps more correctly, how to *live the lack*. To live it by projecting it, and that means using it to recognize and appreciate other things, other people.

If this is the case, then nothing is actually beautiful, as it were, by birth. And it would therefore be wholly inappropriate to demand of *art* that it be beautiful. Ultimately, art wants us to find it beautiful, and if we are unable to do so, then that is our problem, our failure. Yes, the petite bourgeoisie uses art to vent its resentments. And when art slips out of its grasp, it becomes – I don't know what word to use – "theoretical" (that will have to do). And that takes work. And how! But that work is not work on art; rather, it is also, as mentioned above, work on oneself. In order not to shrink from this effort, one must first have respect for oneself. That, however, is precisely what the global petite bourgeoisie does not have. One must feel that one owes oneself something; that is, one must recognize one's lack. And like the Chinese painters, one must afford oneself the luxury of theorizing. The literal meaning of "theory" is "to see." Not to chatter and make pronouncements, as I am doing right now, but to see, to look. To take time. For it takes time to accomplish this change of registers, to shift between the perspective of those who patent plants and that of the young Marcel. "The buttercups grew past numbering on this spot which they had chosen for their games among the grass, standing singly, in couples, in whole companies, yellow as the yolk of eggs, and glowing with an added lustre, I felt, because, being powerless to consummate with my palate the pleasure which the sight of them never failed to give me, I would let it accumulate as my eyes ranged over their

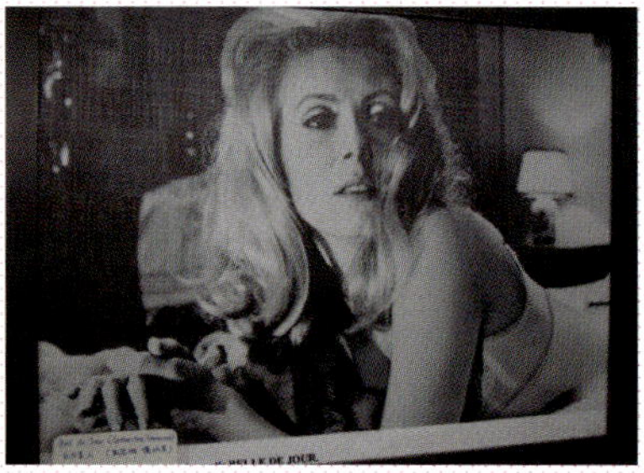

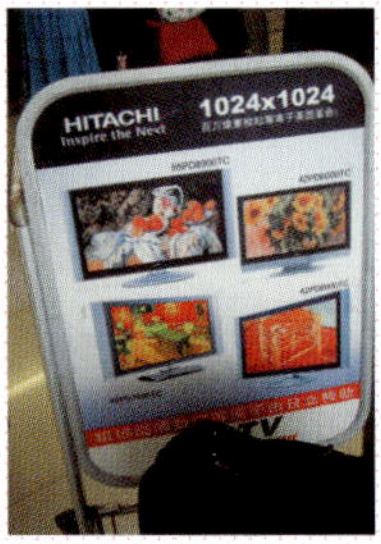

Michel Cluizel: Untitled (left, above) and Orchid (right)

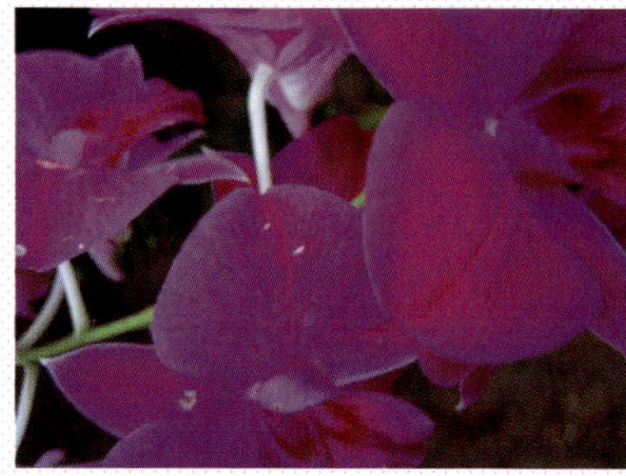

gilded expanse, until it had acquired the strength to create in my mind a fresh example of absolute, useless beauty..."

Useless beauty, that "pleasure" of the "palate" – these are already the first steps on the road to crime, to feeding on and ingesting or assimilating things. For it is here that the regressive and progressive varieties of projection part ways – projection as fixation, as a petit-bourgeois lifeboat on the one hand and, on the other, projection as correspondence with an idea.

It may be that the newly awakened penchant for beauty also corresponds to those political forces which are no longer willing – or never were willing – to entrust their hopes for the future of the planet to capitalist parliamentarism. The funny thing is that these political forces, which come together, for example, at the social forums, have little or nothing to do with beauty. Or am I mistaken? Is it simply that different norms of beauty are at work here, and that one of them enjoins us not to speculate about beauty?

Or perhaps these forces merely have nothing to do with *art*? Why, after all, should art be the privileged vehicle of beauty, when a flowery meadow clearly does just as well? Or is the flowery meadow not a flowery meadow at all, but rather the image of a flowery meadow, an aesthetic construct? Beauty, at any rate, only has a chance if it is able to join with an idea, and today that can only mean an emancipatory political project. The curtain has long since fallen.

This essay was written as a critical review on the occasion of the exhibition *About beauty* at *Haus der Kulturen der Welt in Berlin* (March 18th to May 16th 2005).

1 Hartz IV is the controversial new reduced-rate unemployment benefit in Germany.

ULF POSCHARDT

-------MUTANTS-------

Don't be afraid of the century of self design! The old school of self-design involved fashion, hairstyle, mindset and bloody cosmetic surgery. Courtesy of genetic therapy, 21st century citizens face totally new opportunities, but also risks. The rapid advances made in biotechnology and genetics are enabling an exponentially increasing number of opportunities at ever shorter intervals. But more opportunities also mean greater freedom of choice. And greater freedom means more responsibility.

Bearing this in mind, the mutant as a phenomenon – i.e., the subjectivization of a mutation – is primarily a projection screen for the fears that go hand-in-hand with increased opportunities. Particularly in Germany, a great many of the concerns about progress stem from the responsibility these liberties are known to entail. There is consensus across all political parties in Germany that maintaining the genetic status quo should be grasped as a kind of unspoken norm. This would mean directing a conservative and generally faint-hearted gaze at a *possible* future. But then it would be impossible for the mutant to become the model for fundamental diversity in human ideas on norms and measures. Which makes it so fascinating to see how playfully and at times cheerfully the mutant crops up in contemporary culture: from trivial science fiction comics to the highbrow literature of Michel Houellebecq or the mutant artwork by Thomas Grünfeld, the brothers Jake & Dinos Chapman, Matthew Barney, Milan Knizak or Paul McCarthy.

As civilization advances a growing number of options for human self design are becoming available. Surreal details in the works of Hieronymus Bosch, Goya and Grandville or Dali and Magritte are becoming actual facts. The now famous mouse with the human

Advances in biotechnology and genetics mean man can take the design of his race into his own hands. But they also involve greater responsibility. The mutant is a phenomenon that divides society, stirring up hopes and fears alike. Scientists and medics are equally divided. Many people regard the increase of mutations produced by toxic load as a threat to mankind. Yet genetics researchers are given budgets worth billions as part of affluent society's efforts to end disease and handicaps. Even in a society in which concepts based on origins such as that of the family have virtually ceased to play a role, the uncertain future of mutants gives rise to unease. The majority of people find even a healthy mutant frightening. Comics and films still interpret mutants as monsters that get out of control. Though their skills fascinate us we nonetheless feel threatened by them. What if a mutant disguises itself as happened in

ear on its back is like a painting by Magritte or Max Ernst – come to life. And the photographic agency's image of the said mouse shocked many people to the core. Essentially, mutation no longer belongs to the realm of figures from horror movies and surrealist paintings but is part of our reality, and we must respond to it. The next mutations will affect people.

We must no longer remain what we are born as but have endless opportunities to not only grasp ourselves anew but also to recreate ourselves. Identity can be tinkered and experimented with. In the development novel written at the beginning of the Enlightenment, becoming an adult involved making greater use of your reason. Today, becoming an adult entails accepting full responsibility for your appearance with all the consequences involved. I am what I would like to be.

In an age in which we call into question man as the measure and norm of technology, medicine, metaphysics and religion, science and culture, what is called into question is in danger, and accordingly assumes immense significance. If in the future genetically engineered mutations and mutants become part of everyday life, we will have to think about the past of these human constellations. After all, every present requires a corresponding past.[1] Dialectically and historically the anthropomorphic mutant as a hybrid, transitional creature raises the question about the not-yet-mutated: Kant's question "What is man?" which has shaped discussions in the humanities must be complemented by the question "And what is a mutation of man?". This question is posed alternately in the cultural and scientific domains, two realms that influence each other, though in the 20th century the discursive power of science and medicine tended to sideline the cultural aspects of perception and knowledge. Social "reality" is particularly prone to being shaped by discourse in times in which it forms itself, but that does not make it perceivable or formable.

the movie *Blade Runner*? Yet contemporary graphic artists often have a playful, cheerful approach to mutants. At the beginning of the 20th century Dutch botanist Hugo de Vries discovered that mutations occur by leaps and bounds. If we take as a starting point Luhmann's theory that a system is capable of restabilizing itself, when mutations occur art and culture might be assigned the task of making these leaps tolerable for people. However, society would need to undergo a paradigm change: humanistic thinking would then be deemed a pessimistic, conservative view of society, while liberal positions would shift to the role of revolutionary taboo breakers. And we would have to modify Kant's question: "What is man?" to incorporate the idea of what is feasible, asking instead "What kind of person do I want to be?".

Foucault believed the end of mankind marked a return to the beginnings of philosophy, provided the humanities were woken from their anthropological slumber and accepted the insecurity of postulating non-anthropocentric views and perspectives. In Kant's early 19th century anthropology, becoming human was coupled with self-knowledge and as a consequence the question about what constitutes a human-being was replaced by thinking about the Self. Seen in the context of Kant's anthropological thought, the concept that emerged in the course of the 20th century about the non–Self, the subconscious, multiplicities and irrationality could herald the beginning of the end of mankind. Or it is the first step towards designing a new human-being, who is far removed from an inexorably advancing "egoism"(Kant).[2] For this to happen a few more threads would have to be removed from the edge and filling of our culture and analyzed, the threads Foucault described as part of the dark but solid fabric of our experiences.[3] As a result, the blindness of designing a new negative foil can only be reduced to some extent, but it can at least be understood in parts as blindness.

The discovery of mutations and their design serves to intensify man's quasi-naturalness. Since mutation was seen as forcing man's civilization, the idea – based on nature's forces – that man is by nature artificial, as Hellmuth Plessner put it, became a central notion. But it is no longer the consciousness and emotional system that are the manifestation of this artificiality; the flesh itself is seen to be created from the genome's smallest modules. The dynamism of change shifts from changes in the environment to changes in the essence of the Self. As such, it is our own wishes and nightmares we are responding to when we accept the mutant as a figure in our consciousness. The treatises and language of exclusion produced by a defensive attitude are symptoms of the vigilance and agitation innate to the cultural negative foil. In the movie *X-Men* strict monitoring and control are incorporated into the prison created for the "evil" mutant Magneto. Created

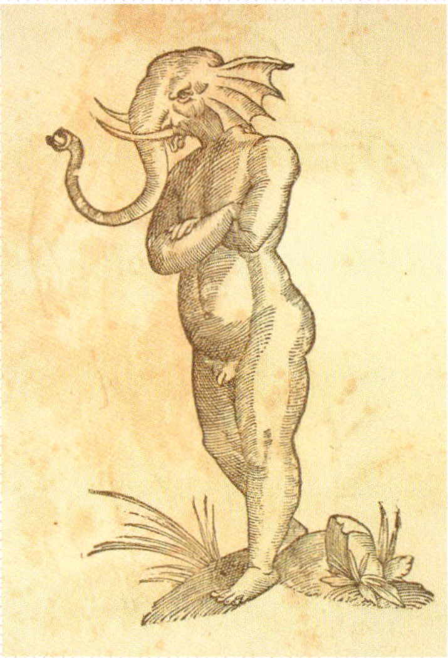

In his "Historia Monstrorum" mathematician and physician Ulisse Aldrovandi (1522–1605) presented not only malformed humans, animals and plants, but also a series of fabled creatures. The "Puer capite elephantine" (boy with an elephant's head) was purportedly born in Hungary, so Aldrovandi claimed.

at the movie's close this state represents the Happy End. For the sciences and medicine the increase of mutations produced by the higher toxic pollution of water, air and food not only poses a threat to the gene pool of the current generation but also future ones with unforeseeable, catastrophic consequences "for which we should feel responsible."[4] It is a moot point who is meant by this "we": science, all global citizens or all polluters of the environment? A feeling of guilt and responsibility towards the victims of mutation has led scientists to search for the causes of spontaneous mutations and thus identify those responsible for alterations in genetic material. This development has led to mutation being associated with illness, and the consequences of civilization are viewed as the cause behind such mutations. The need for good health not only implies that people must have a sound physical constitution, but also a healthy lifestyle and by dint a healthy environment. Non-mutation demands a totality of health. Not only is cigarette consumption made responsible for changes in genetic material but also the atomic radiation caused by the bombing of Hiroshima and Nagasaki or accidents in nuclear power stations resulting in a high incidence of deformed children being born in the immediate vicinity. In other words, in everyday usage mutations are associated both semantically and emotionally with diseases similar to cancer, which can be understood as abnormal organic growth. At the end of the 1990s, researchers into genetics attempted to adopt a sober attitude towards mutations and mutants; after Darwinism the importance of mutations have always been under- or overestimated. The fact that controlled mutations can be produced in the lab has helped focus attention on the biotechnological implications of the altered genome. As a result, contemporary genetics is eager to describe in as factual, non-sensational manner as possible the phenomena caused by mutations but also the phenomena and conditions that cause mutations. The emphasis is on clarifying the causal connections.

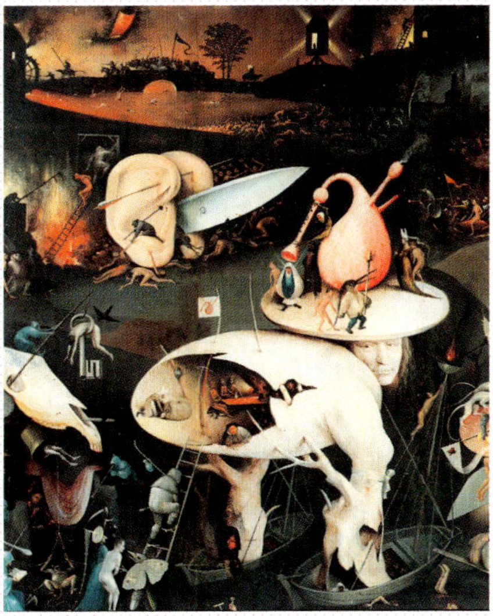

In the panel "Musical Hell" from his triptych "Garden of Earthly Delights" (of which a section can be seen here), Hieronymus Bosch (1450–1516) presented a world full of real and imagined horrors; torture and sexual obsessions are shown either directly or symbolically. In the center of the image populated by fabled beings and symbolic figures stands the so-called Tree-Man looking contemplatively out at the viewer.

It is difficult to predict what the consequences for evolution will be. Increasingly, the progress of research is no longer being determined by scientists who perform analyses and investigate reactions but those who define themselves by the genetic material they design and create. This branch of science – financed by the biotechnology industry – has access to research funds that are unusually generous even compared to those received by other scientists. Altering the human standard is an undertaking driven by a budget worth billions and the expectations of the welfare society are accordingly high; above all people hope to see the end of diseases and handicaps while suppressing the issue of side effects. The global citizen's view would seem to be based on the wish that everything else should remain as it is: everything should become better, and nothing worse. The use of the word mutant as a term of abuse[5] indicates that through the 1990s the idea of mutation still provokes fear and loathing. The mutant has come to replace the diseased person, a new phenomena different from the definable, healthy or rather non-mutated subject and its discursive impact. Moreover, the fact that the concept of healthy is intertwined both semantically and logically with the idea of normal implies the potential for possible exclusions. Mutants are created by physicians and biotechnology researchers and unlike the ill patient are not returned to a state of health. Even the good mutation, i.e., the healthier and better person, is a frightening prospect for us, and we require genetically manipulated food to be labelled accordingly, even if it allegedly tastes better. The bad mutation as a mysterious, puzzling, dangerous being leaves unanswered questions about its origins: is it a cross with an alien or a wonder? The certainty enshrined in *mater semper certam* no longer exists. What makes the mutant so mysterious is the uncertainty about its origins, traditions and descent. It is this uncertainty that places it in the ranks of monsters, deformed creatures and freaks. This fear of the contamination of the concept of relationship as it exists in the family also functions in societies which no longer lay any store by such concepts.

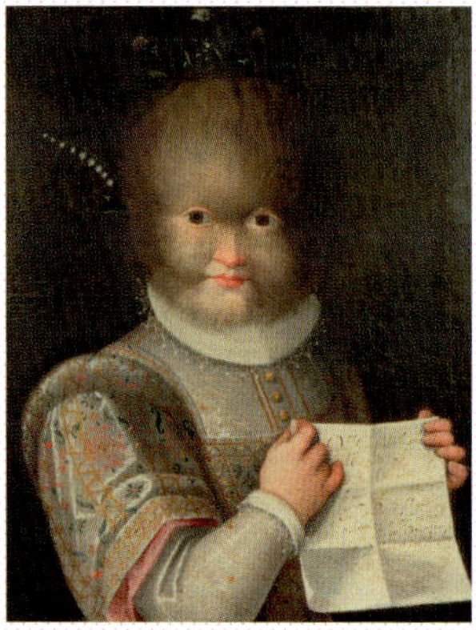

In 1585, painter Lavinia Fontana produced the portrait of 11-year-old Tognina Gonsalvus, the "fur girl" from Antwerp. As a living curio, Tognina and her likewise very hirsute father were taken to the Court of the Duchess of Parma, where they were paraded as monsters and marvels.

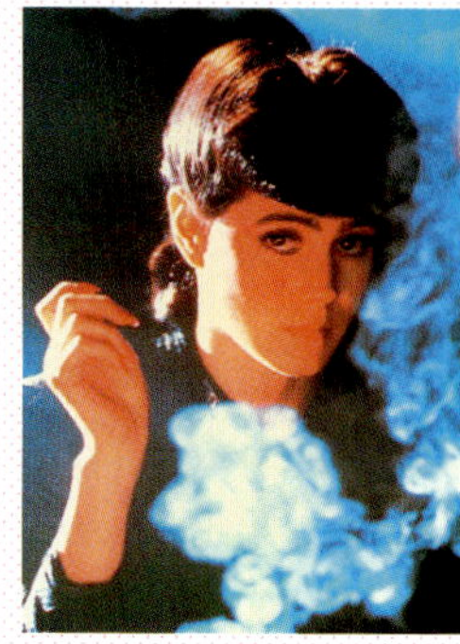

Ridley Scott's film "Blade Runner" (1982) paints a somber image of a no-future vision. The long-winded cult film plays in a world in which dangerous androids called "Replicants" live alongside humans. Can a difference be discerned between humans and androids? And to which of the worlds does the mysterious Rachel belong?

The grotesque appearance of mutants torpedo our longing for the idyllic. They are a thorn in the sensitive and frightened emotional flesh of those who wish for normality within the constancy of continual life improvements but without having to take leave of the old benefits of contemplation to which they have grown attached. More dangerous than the grotesque-looking mutants are those that cannot be detected as such from their appearance. Those mechanisms of the gaze and perception that Foucault described as decisive for discourse and thus for definition, come up against their limitations. Just how traumatic not being able to trust your vision is for a visual culture can be seen in the movie *Blade Runner*, in which increasingly sophisticated methods are developed to identify the human clones. However, the point is that we may not be able to trust the detector himself, who could also be one of the highly-advanced clones.

Mutants are frightening apparitions. And they seem to embody all people's fears about the disturbing possibilities of genetics experiments, which feature so prominently in the media and politics. The non-linear rampant development of the body and intellect out of sync with the course of evolution pierces the unchanging anthropometric factors, which until now have withstood the acceleration of our habitats, changes in socialization and upheavals in civilization. This rampant development is the potential intervention in what we refer to as our own "flesh and blood".

Movies and comics such as *X-Men* address man's fear about the threat mutants pose to the human race. In the former, a special Senate Committee is appointed to deal with the issue. Mutant expert Jean Grey, who – unknown to everyone – is herself a mutant – explains that mutants are those people who make the most of their DNA potential in order to accelerate evolution. This would put an end to the concept of evolution as a slow process. However, politicians demand all mutants be registered and monitored. This conflict is a central theme of the movie. As the evil guy Magneto, a resident of the Jew-

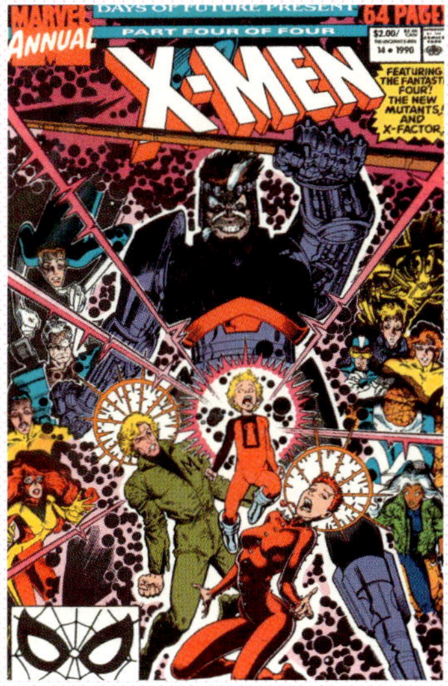

The comic superheroes "X-Men" launched back in the early 1960s. Yet it was the redesign of the characters in the mid-1970s that brought their enduring success. Good mutants with superpowers protect humans against egotistical evil mutants. The comic series has since spawned a popular series of movies.

ish ghetto in Warsaw, himself lost his parents because they belonged to a "foreign" race he entertains no hopes for a peaceful coexistence between mutants and non–mutants. His prediction is that either mutants will rule the world or they will be destroyed by humans.

In the early 20th century, the figure of the mutant is increasingly present in art, but also in films, literature and theory. As mutants are still fictional figures, analyzing mutants means analyzing the media in which these figures occur. The presentation of the figure "mutant" in the culture industry tends to be completely different from its portrayal in the fine arts, philosophy and contemporary theory. In the mainstream it is a descendent of the monster, a threat with mysterious powers, either vulnerable and good or hurt once and subsequently evil forever (e.g. *X-Men*). Mutants' appearance and genealogies are cited in Pop culture, in comics and in science-fiction movies and books, while in the fine arts and in design the work of art or the object is experienced as a mutant. In other words, while in Pop culture the debate takes place in the most direct and obvious form of the subject, artists and designers who see themselves as engaged in serious art generate the kind of forms that would seem to be derived from the processes of mutation. Given the extensive networking with the canon of subjects addressed by Pop culture and its images and myth-spewing energy, the mediated reflection on mutation and mutants often appears as a force-field in art that considers itself autonomous. Carried away by a fascination for the sciences, the mimesis of the arts has devoted itself to the poetics of genetics as a leitmotif of its own creation. Art mutates with the sciences at the level of self-description but also the levels of production, reflection and on the subject level. Contemporary theory takes a similar approach when playing musically with concepts and terms.

Given the possibilities of an evolution pushed by technology and environmental influences, but which questions its own concept, as a construction the mutant is – as intimated – a reflection of human hopes and fears. The alien and the monster represent in

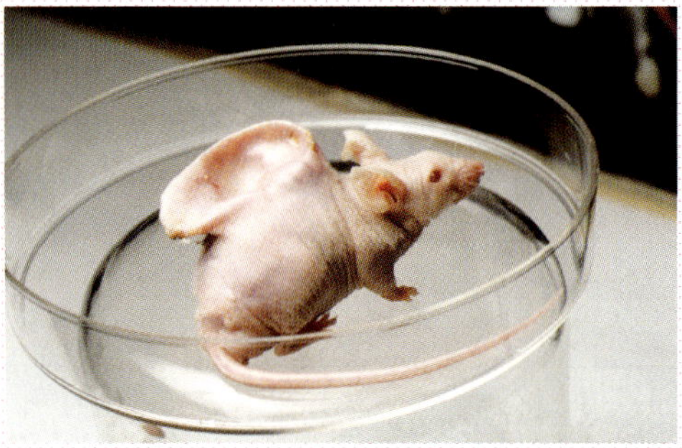

Using biologically degradable plastic, enriched with cartilage cells from a cow, Charles Vacanti, an anesthetist from Boston, shaped a human ear and implanted it on a mouse's back. Gradually the growing cartilage replaced the plastic. As a method, tissue engineering is a milestone in transplantation research. The mouse with the ear on its back became a source of public debate, however, as the popular symbol of the horrors of untrammeled genetic engineering.

the form of frightening antitheses man's attempt to define and attain certainty about what constitutes his own self even though insecurity about this very matter has long since pushed us to alter our views on this topic.

What makes mutation threatening is the fact that it evolves from its own organism, which – this is what constitutes the threat – could itself become alien, without being able to separate itself from this foreign aspect. Mutation is the exaggeration of one's own development, which you can no longer understand and influence by applying old concepts. The conquest of your own body is no longer the result of disciplining or controlling the Self, but the body's controlled and modified ontogenesis, which can have a much more material influence on the evolution of the subject and the design of the subjectivization than socialization.

Simultaneously, science fiction and anthropo–fiction claim at least potentially that scientific design dynamism could be inspired by cultural dispositives. The poetic premonitions about future anthropogenic life forms will be selected in accordance with the needs of the environment and its consequent functionality. If you accept Luhmann's theory about the system's urge to restabilize itself, then, against the background of this potential re-stabilization, selection would be capable of allowing a larger variety of inconstant mutations, in order to provoke re-stabilizing mechanisms into incorporating these mutations into the overall system. The potential wildness of the mutations, in particular the anthropogenic mutants, would be influenced by socially and culturally co–mutated or better evolutionized versions of some new species of man.

Let us reiterate: If organisms mutate it could be the task of cultural concepts, traditions and ideas to make these leaps and chaotic states tolerable, because culture could largely continue to evolve constantly within the liberty of poetry, far away from the touchstones of reality. Culture would then tend to be more of a band of continuity than

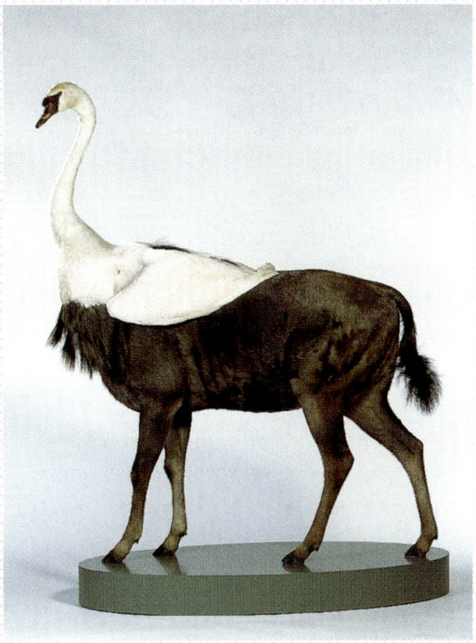

With his series of chimera entitled "Misfits" Thomas Grünfeld highlights genetic engineering and the cheerful science. The picture shows a taxidermized cross between a gnu and a swan (2001).

›
Sympathetic and disturbing:
"Spaghetti Man" by Paul McCarthy, 1993

a motor of arbitrariness. However, such a description is new and at variance with the prevailing concept of art and the artist's self-image; after all, the role of art and the artist producing it has always been subversive, belligerent, driven by a passion for the new. The strong interaction between scientific discoveries and theories on the one hand, and cultural, humanities reflections on the other, led to art and culture drifting in the 19th century and more strongly in the 20th century into an anthropocentric identity and a diffuse force. Culture effectively created protective areas: simplifying in the form of pigeonholes, soberly in the form of lifeworlds, which could make the shifts and dynamics of the new age tolerable – since they could be reflected on.

Possibly, culture is generally more called upon as a restabilizing force when the sciences and technology produce wild variations, which the system must then select. We can only speculate as to whether culture will accept this task or whether it has (as a ruse of history, as it were) already done so before it could realize how important this commission would be. If culture and science act as a strong room and archive of our concepts and traditions it is possible to assume markedly more social responsibility without renouncing the dogma of the autonomy of the arts. If science contents itself with viewing the present as the presence of what is past, in order to shape the future, culture can secure the future presence of what is past.

If that is the case, then theory as the aesthetic of semblance will as the brief and memory of a design duty return once again to unusual paths. Thanks to its memory function, culture can make the future tolerable – even if this tolerable future life must be thought of in nomad terms in a retroactive vitalization of past. The amassed past as a resource will be examined by cultural sub-systems and the arts as to its potential long-term use. The selection mechanism of culture functions in both directions: as active forgetting in the form of shaping the past but also as an active designing in the guise of visions for

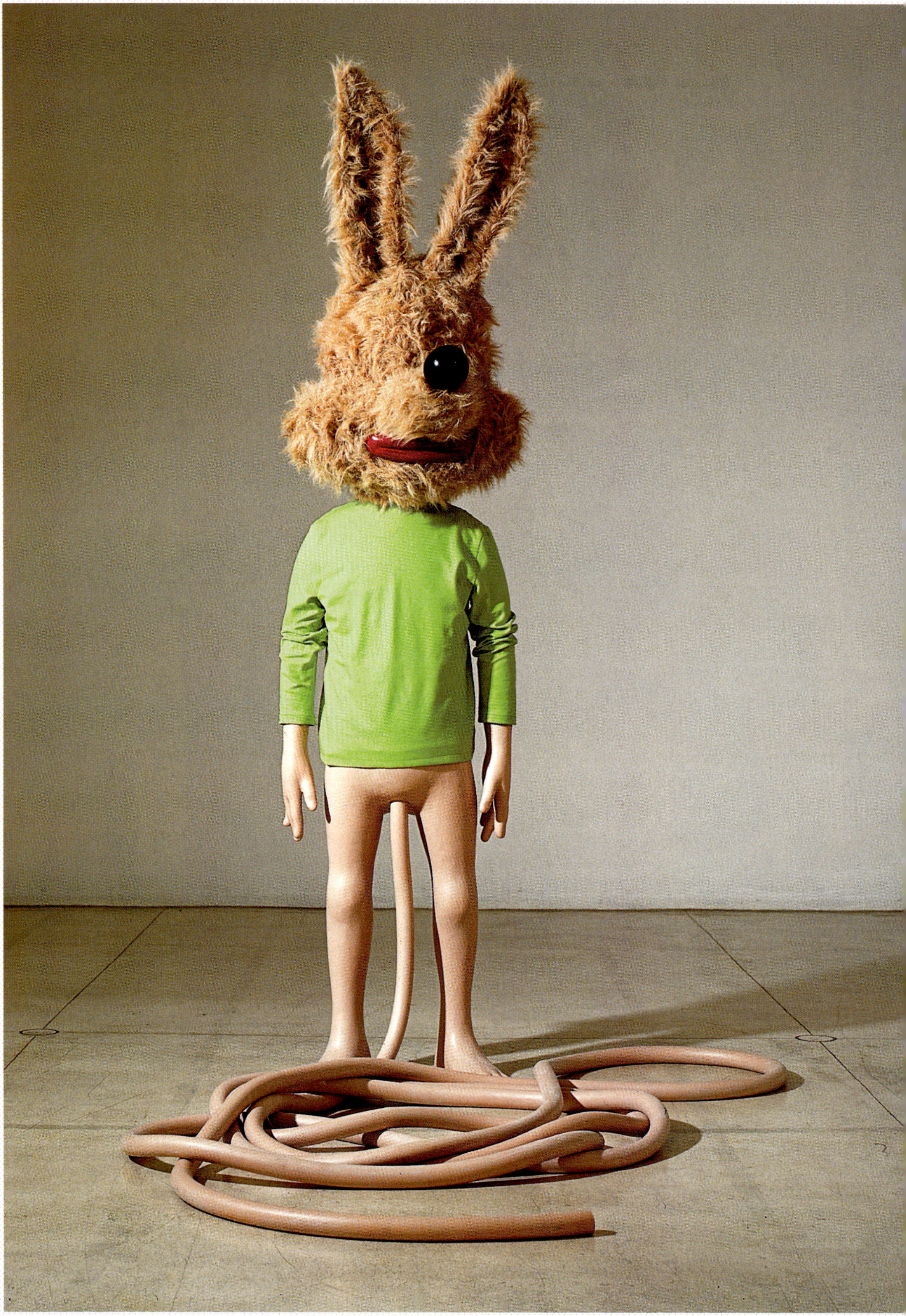

the future. The band of continuity created in the process offers stability, without having to be binding.[6] Whether culture and the arts allow themselves to be pushed into the defensive by the sciences and medicine or not, will determine their significance.

When Amsterdam botanist Hugo de Vries presented his mutation theory in 1901, he wanted to liberate the theory of the genesis of the species from the constraints of its existence as a comparative science. Contrary to the assumption that the genesis of the species could not be directly observed or subjected to experiment, he sought to prove that species emerge “by fits and starts” and that these fits and starts can be observed like any other physiological process. Perceptibility should be derived from the speed of the processes that were to be observed. He did not share the view that species of animals and plants emerged from one another “very gradually”. That said, such speed would make it impossible to perceive the emergence of a new species, even if a researcher were to devote his entire life to the endeavor. As such the mutation theory was also the manifestation of scientific ambition and self-confidence.

Perceiving the dynamism of genesis and the processes involved compels researchers like de Vries to question the scientist’s role. If the processes occur sporadically, then it would not only be possible to render mutation observable but also to experiment with these processes, i.e., to act as creator. De Vries performs the step from perceptibility to formability in one. Using the classic methods of biology and palaeontology it would be possible to reveal the evolution of new species and their laws. The background for the perceptibility of discontinuities and leaps “without transitions” is variability, which can take place both continually and in leaps and bounds. The relationship of variability and mutability as a relationship of continual or intermittent evolution of species means for the botanic discourse that it is possible to register both. Identifying the difference between the two is both method and outcome in one.

The variations created by mutation give rise to new units. According to de Vries this new unit “signifies” a mutation. The new unit is the prerequisite for a new variability, after which it can develop. After proving the difference to be both method and outcome, the definition of a new unit amounts to an entelechy of mutation, which allows the development of a so-called mutation theory. Having isolated a unit, a discursive regularity is prepared with which science in the face of a dynamic phenomenon can secure itself a stable, albeit adaptable standpoint. The gaze is directed towards the future. “The knowledge of the laws of mutation,” writes de Vries, “will probably lead to the production of artificial and arbitrary mutations and consequently give rise to totally new characteristics in plants and animals. And in the same manner that you can use the selection procedure to create refined, more profitable and more attractive breeding races, it will perhaps be possible in future through the mastery of mutations to continually produce improved types of cultivated plants and animals.”[7]

This prophecy once belittled by scientists is becoming ever more likely with each advance in biogenetics. Culture is divided into two camps: that of the admonishers and brakemen, who insist on ethical and anthropological standards and seek to stop, slow down or prohibit the futuristic vein of research in order to eliminate all signifying units, and the camp of those who wish to stand at the forefront of those innovations with which humanity is rushing towards its new future. What is interesting about this is the way that in particular humanist thinking from leftist traditions is becoming increasingly conservative and pessimistic towards civilization, while liberal positions slip into the role of revolutionaries who break taboos.

This paradigm change is clearly evident in culture. In the digital maelstrom of the loss of identity culture is losing its old familiar face in the service of its mimesis. Murphing as the manifestation of mutation, the digital collage as the circuit diagram of desired mutations, the sampled sound as an audio experience of a world that has fallen out of all certainties and uniformities. In radical thinking culture has always been a tool of mutation: it deeply alters the life and thinking of people. The world-changing force of idealism has banished the creature man with his gift for reason into the maelstrom of modernization, at the end of which lurks only the destruction of the alteration processes.

It is not just a case of supplementing Kant's four questions by the question mentioned previously "What is the mutation of a person?". In future these essential philosophical questions will have be supplemented by a subject-centered feasibility question: What kind of person do I want to be? The performative will call the normative into question and seek to replace it. The luxury of being able to answer this question as honestly as possible has always been a feature of social privilege and power and this will be all the more so in future. The happy mutant is a person rich in self-design. The reversibility of self-models makes life into a long, turbulent continuum – if this is desired. Which ought to put an end to man's boredom with himself.

The mutant is the radicalized figure of those novels of education, which since the Enlightenment have primarily illustrated man's genesis and development as an intellectual process. The geneticization of entelechy also signifies a decentralization of the body, which no longer necessarily sees all authority as being located in the brain. I mutate, ergo I am.

And: are you still afraid?

1 Luhmann, Niklas: *Die Gesellschaft der Gesellschaft* (Frankfurt am Main, 1997), p. 445

2 Kant, Immanuel: "Anthropologie in pragmatischer Hinsicht", in: Kant, I.: *Werkausgabe* vol. XII. (Frankfurt am Main, 1977), pp. 407–409

3 Foucault, Michel, *The Birth of the* Clinic. An Archeology of Medical Perception (French: *Naissance de la clinique: une archeologie du regard medical*, Paris 1963) (London, 1973)

4 Obe, Günter: *Mutations in man* (Berlin, 1984), p. V

5 Poschardt, Ulf: *DJ Culture* (Hamburg, 1995), p. 117ff.

6 Luhmann postulates that the memory loses its function of providing clues because culture has lost its certainty about where it is going. In the context of a political and ethical discussion all the parameters shift: cultivating tradition once a typical conservative role seems to have landed on the side of a humanist ideology once defined as leftist, while actively forgetting the past and focusing on the future is encouraged by the political apologists of economic liberalism. In other words, depending on their interests in the *all over* of culture one's own is re-discovered and encouraged.??

7 de Vries, Hugo: *Die Mutationen und die Mutationsperioden bei der Entstehung der Arten* (Leipzig, 1901), p. V

FRANK EDENHOFER

------- ON THE MATERIALITY OF LIFE -------

> "Medical science's attitude has become mechanical,
> its goal defined as the identification of the identification of
> an organism's physics. It has proven that life itself is
> but the expression of a sum of phenomena, each of which
> follows the ordinary physical and chemical laws."
> *Rudolf Virchow (1821–1902)*

About 150 years ago, Rudolf Virchow (1821–1902) developed the concept of cellular pathology. Although taken for granted nowadays, the underlying idea was highly controversial at the time: Virchow postulated that disease can be caused by dysfunctional cells in the body. This was totally opposed to humoral pathology, or the "four humors theory", which maintained that the fluids (humors) in the human body are responsible for illness. A balanced mix (*eucrasia*) of the four cardinal humors – blood, phlegm, yellow bile and black bile – supposedly resulted in health, whereas an imbalance (*dyscrasia*) was believed to cause illness. Humoral theory dominated methods of medical treatment well into the 19th century and held that a doctor's healing mission consisted of re-establishing a state of eucrasia by means of a suitable diet, medicinal remedies and, if necessary, surgical operations. In opposition to such teachings, Virchow's reductionist concept of cellular pathology claimed that it is not the entire body that becomes diseased but only individual cells or groups of cells. A provocative hypothesis in its day: the advocates of humoral pathology believed that nature is perfect and incapable of error. Virchow's cellular

Approximately 150 years ago, Rudolf Virchow's theory that the dysfunction of individual cells or groups of cells can cause disease was a highly provocative concept. It opposed the then commonly held "four humors theory", according to which good health depended on a balanced mix of the four cardinal humors – blood, phlegm, yellow bile and black bile. Vierchow's concept was the basis for a dynamic development in modern medicine. The better nature's cell design was understood, the more it was possible to attribute morbid changes to dysfunctional cells. Nevertheless, for a long time the repair or even replacement of defective cells remained nothing more than a vision of cellular pathology. Today, says Edenhofer, we may well be at the threshold of a new development in cell therapy: specific cell design. There are two strategies for this approach: the exploitation of cells' natural potential to regenerate

pathology cemented the paradigmatic end of a development that had been slowly but constantly progressing for centuries, a development in the way man perceived living things and the dysfunction of the same. It was nourished by empirical observation and the resulting rational knowledge of the materiality of life, manifested in the mutual relationships between bodies and substances. The start of this development was heralded by English naturalist Robert Hooke (1635–1703), the first person ever to record his observations of a new, hitherto unknown microscopic world. In the book he published in 1665, *Micrographia or some physiological descriptions of minute bodies by magnifying glasses*, Hooke described in detail his microscopic observations of plant tissue, such as that of a thin slice of cork. It was he who coined the term "cell" (from *cellula*, a small room), in allusion to the tiny chambers occupied by monks. In the second half of the 17th century, microscopy made some dramatic advances – driven both by technical progress in the manufacture of high-quality optical devices like glass lenses and scientists' urge to gaze on unknown, never-before-seen worlds. Although the credit for inventing the microscope cannot be clearly attributed to any single person, it is interesting to note that two astronomers, Galileo Galilei and Johannes Kepler, are generally thought to have been major contributors to its development in the early 17th century. With their telescopes, they studied the little-known world of space, the moon, the planets and the sun – or, to put it another way, the world of the huge, the distant, the external. They then converted this optical instrument into a microscope, enabling them to observe the equally unexplored world of the small, the near, the internal – and thus the materiality of matter, especially that of life. Dutchman Antonie van Leeuwenhoek (1632–1723), originally a draper from Delft, played a crucial role in driving this development forward by keeping extensive records of his microscopic observations and achieving a hitherto unparalleled perfection in the manufacture of optical lenses. In herbal brews, drops of water and dental plaque,

or the creation of a custom-made cell with the desired properties. Stem cells are immature cells that develop into mature, fully functional cells and have an almost indefinite ability to reproduce. Alternatively, even less mature cells can be used, the so-called pluripotent cells found at the very start of an organism's development. Because they are not limited in terms of what they can develop into and can potentially turn into any of the different cell types found in the body, embryonic stem cells are particularly suited to such purposes. Their development can be specifically controlled by adding hormones, growth factors and other extracellular signals. Despite the initial, encouraging results with transplanted cells, therapeutic use for cell design from stem cells is still a long way off and involves a host of unpredictable factors. The problem of how individually cultivated cells that are then integrated via trans-

he discovered tiny animated bodies that he described as *animalculi*, or "little animals". He also ascertained that human tissue or fluids such as hair, blood, saliva and semen contain the same type of bodies. The realisation that these little bodies were to be found virtually everywhere, in a diversity and number never before seen, gradually became accepted: wherever there is life, there are cells. It was evident that there are living organisms such as the *animalculi* that consist of a single cell. However, the structure and especially the origin of more complex life forms remained unclear and the subject of intense scientific debate for a long time after Leeuwenhoek. How do such cells come into being, how does a person develop? In the late 18th century, Swiss naturalist Charles Bonnet (1720–1793) developed the idea that female organisms contain all future generations within themselves, in miniaturised form. After its birth, he believed, the development of the fully grown organism consisted entirely in the enlargement of this tiny copy. Such *homunculi* (as they were called in the case of humans) dominated the self-image of the natural sciences and philosophy for a long time afterwards. Johann Wolfgang von Goethe (1749–1832) gave shape and form to such a *homunculus* in the second part of Faust, by having Wagner bring him to life virtually out of thin air. Up until the mid-19th century, this kind of *generatio spontanea* remained the predominant opinion, not only amongst poets but amongst natural scientists as well. In 1838, Matthias Jacob Schleiden (1804–1881) formulated the then extremely daring hypothesis that all plants consist of cells. Soon afterwards Theodor Schwann (1810–1882) expanded this hypothesis to include animals as well, attempting to prove his theories by keeping detailed records. Nevertheless, it was to be several decades before the cellular theory postulated by Schleiden and Schwann was generally acknowledged. The advent of considerably improved microscopes in the 19th century and the use of special dyeing methods for certain cell components contributed to the widespread acceptance of their theories. In addition, the open

plantation can functionally reconstitute the mature cell aggregate of a complete organ will also require intense research. In addition to the biological limitations, however, Edenhofer also refers to ethical concerns about using embryos for the production of embryonic stem cells. Current research approaches are concentrating on alternative ways of producing pluripotent stem cells. There are some initial indications that reprogramming already developed cells into immature stem cells could be a feasible course of action. But nothing is possible without cells – there has to be at least one cell at the beginning.

question as to the source or origin of cells was becoming the subject of more intense enquiry. French microbiologist Louis Pasteur (1822–1895) conclusively refuted the idea that cells can form spontaneously out of inanimate material, thus burying the theory of *generatio spontanea* for good. Finally, based on these learnings, Rudolf Virchow developed the theory that life can only be created by direct succession. He further concluded that every cell has to be created from another (*omnis cellula e cellula*). Wherever there is life there are cells, and no cell can be created out of nothing. Virchow also postulated that pathological changes in the body are due to dysfunctional cells. Accordingly, disease is a phenomenon of life and inextricably linked with it. More than any other theoretical approach before it, the concept of cellular pathology symbolises the transition of medicine and the life sciences into the enlightened modern age.

THE CELL – THE SMALLEST UNIT OF LIFE

The human organism is made up of 10^{13}–10^{14} cells. An astronomical number that is almost impossible to grasp: if we could gather together the populations of 10,000 replicas of planet earth, we might have an idea of just how many cells each and every one of us carries in his body. Not all cells are the same – there are about 220 different kinds (cell types), specialised in form and function to perform individual tasks. Some fulfil their purpose in connected cell aggregates (tissue or organs), such as in the liver or skin, whilst others, such as blood cells, go about their work alone. Depending on their function, cells differ considerably in shape and size, from the spherical, approx. 5-micrometre blood cells all the way to fibre-like neurons, the axons of which can measure up to one metre long. On average, a cell measures 20 micrometres, i.e. 50 of them in a closely packed row amount to just 1 millimetre. In chemical terms, cells consist of a series of small organic molecules, chain-like giant molecules, namely polymeric sugar, fat or protein chains,

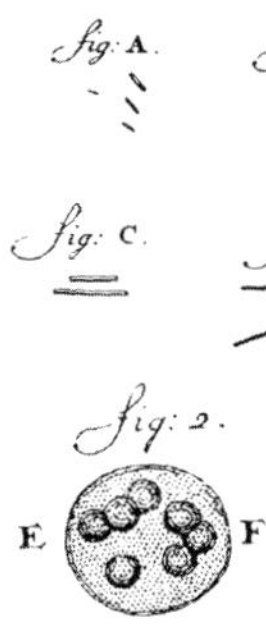

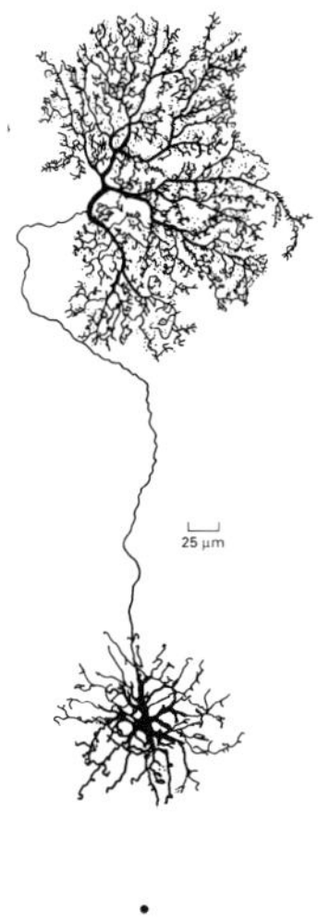

Dutchman van Leeuwenhoek (1632–1723) made his own microscopes to study drops of water and various bodily fluids, finding them to contain tiny living animals, or "animalculi". He conscientiously documented his discoveries in detailed drawings.

Cells, the smallest units of life, can take many forms. The human body is thought to be made up of over 200 different types of cell. Two particularly striking examples: the small, spherical blood cells and neurons (above), the fibrous ends of which (axons) can measure up to a metre long.

nucleic acid and lots of water. It almost defies belief that the incredible diversity of life forms and the functions of a vast array of different cells, as well as the ability to reproduce, can stem from such apparently simple ingredients. The secret lies in their complexity and the functional, mutual relationships between the substances. Each cell is a miracle of biochemistry that puts the accomplishments of human engineering well and truly in the shade. Imagine having to assemble more than 500,000 individual parts into a functioning micro-machine that can simultaneously carry out hundreds of thousands of reactions and thus engender such vital functions as resistance to a pathogen. The whole thing would have to be shrunk down to a size of approximately 20 micrometres and, on top of that, possess the extraordinary ability to reproduce itself. In order to accomplish this extremely demanding task, a cell contains nothing that does not serve a specific purpose, each substance has its own important function. The design of the cellular substances, which has taken millions of years to evolve and perfect, plays a crucial role in all this.

THE LOCK-AND-KEY PRINCIPLE – THE IMPORTANCE OF COMPLEMENTARY FORMS

Almost all cells have a series of structural features in common. A viscous, oil-like layer, the so-called cell membrane, separates the cell from its surroundings and thus acts as a barrier. The genetic information for the structure and function of the cell is stored in the nucleus. Between the two is cytoplasm, an egg-white-like mass in which most of the cell's activities occur. In addition, all cells contain structures bounded by membranes and embedded in the cytoplasm. The mitochondria, in which biochemically controlled, "cold" combustion takes place, are some of the most important of these so-called organelles. These structures provide the cell with energy by burning (oxidising) energy-

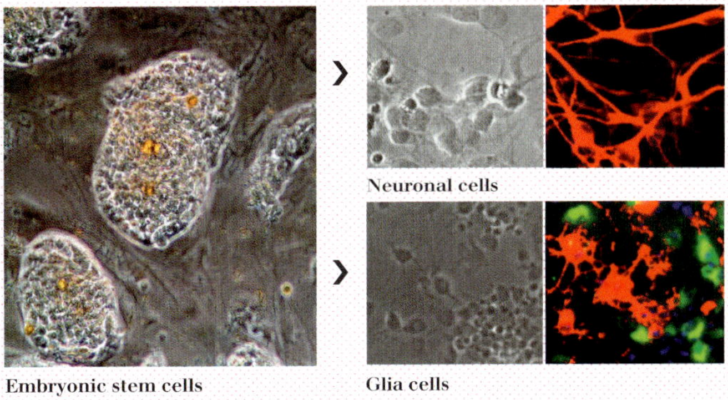

Deliberate cell design by manipulating stem cells. Embryonic stem cells from a mouse (left) form colonies in a plastic dish (note large group of cells at centre). Embryonic stem cells can reproduce almost indefinitely and develop into any type of cell, such as neuronal cells and glia cells. Theoretically, this deliberate differentiation can be used for the therapeutic replacement of any cell type.

rich organic substances such as sugar. If you look at a cell under a microscope, you essentially only see protein. Apart from water, proteins constitute the biggest part of the cell. Besides acting as building materials, they perform almost all the cell functions as well. Embedded in the membrane, which is impermeable to most substances, there are for instance pump and channel proteins. Other membrane proteins function as chemical antennae tuned to the outside, picking up external signals and passing them on to the inside. Within the cell, these signals are often conveyed by messenger proteins, which might bring about cell growth or division or even destroy the cell. Other proteins act as highly specialised nano-machines that transport substances or entire cells, cut up nucleic acids, help produce proteins and so on. It is thanks to their extremely flexible and versatile structure that proteins can produce such a broad spectrum of effects. Nature uses the exceptional malleability of proteins in the same way we can use a piece of clay to create forms of any shape and size to fulfil a variety of purposes.

The “lock-and-key” principle described by biochemist Emil Fischer (1852–1919) is an extremely vivid symbol of the crucial role played by the molecular design of proteins in biological processes. Emil Fischer put forward the hypothesis that there has to be a fit between two complementary forms in order for them to fulfil certain biological functions. He originally formulated this correlation to describe the specific binding of an enzyme to its substrate. An enzyme is a protein with the properties of a biochemical catalyst. Without such biological catalysts, no life would be possible, since most of the metabolic reactions within the cell would either be immensely slow or could only occur under intolerable conditions such as extremely high pressure or temperatures. According to the lock-and-key principle, a biochemical reaction is facilitated by the reacting partner being brought into spatial proximity through a matching structural pocket in the enzyme. Likewise, the effect of hormones on cells is passed on by binding them to structurally

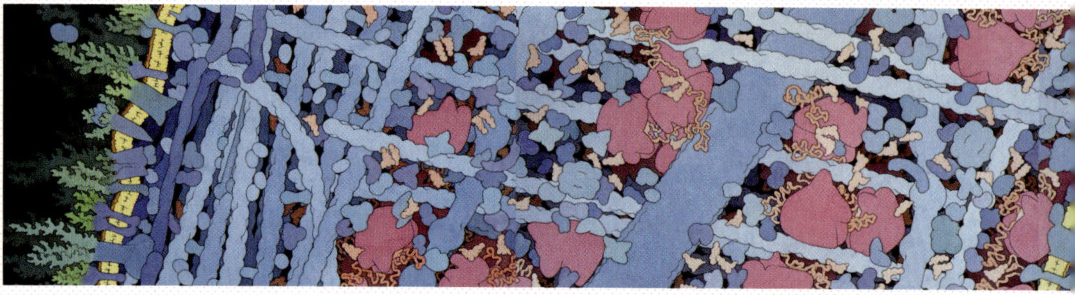

A panoramic view of the inside of a body cell, from the cell surface (left) all the way through to the nucleus (p.96). Each cell is a biochemical-technical microcosm, packed full of macro-molecules such asproteins (in blue), nucleic acids (DNA in red, RNA in orange), fats (yellow) and sugars (green). They include nano-machines, pumps and channels, as well as protein factories (ribosomes, magenta). At the centre of the artist's depiction is a cell power plant or mitochondrion (folded membrane structure).

matching receptor molecules on the surface of the cell. On their surface, all cells in cell aggregates exhibit structures and counter-structures that are important for communication between the cells and thus lead to structural and functional unity. The lock-and-key principle also plays a crucial role in the recognition of "self" and "non-self". The human immune system is essentially based on the complex interaction between extremely diverse structural and counter-structural bonds. This involves the recognition of so-called antigens to pathogens by specialised molecules, the antibodies. Together they form antibody-pathogen complexes, which are rendered harmless by specialised immune cells. The immune system can also recognise endogenous cells that have already been infected because they exhibit telltale particles of alien proteins on their surface.

A NEW AGE: MOLECULAR MODELLING AND CELL DESIGN

Virchow's cellular pathology laid the foundations for a dynamic development in modern medicine and the life sciences. The better nature's cell design was understood, the more morbid changes could be traced to cellular dysfunction. The development of therapeutic methods focused on compensating for the loss of function. Today we may well be on the threshold of a new leap forward in cell therapy: targeted cell design. Up until now, diseases caused by the loss of important cells or cell functions, such as Parkinson's disease, muscular dystrophy or diabetes, have largely been treated indirectly. The diabetes patient is for instance given insulin, the same substance produced by healthy cells. Although this therapy has a satisfactory effect, it does not completely or permanently replace the loss of cellular function. For a long time, the repair or even replacement of the defective cells was nothing more than a vision.

There are two strategies for this therapeutic approach: the exploitation of cells' natural potential to regenerate or the creation of a custom-made cell with the desired

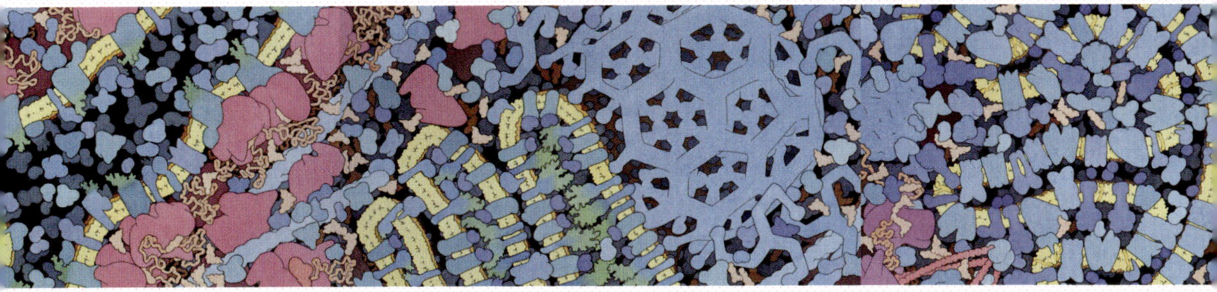

properties. Unfortunately, cells are usually limited in terms of their plasticity: once fully developed, they usually keep their properties. A neuron remains a neuron, a liver cell remains a liver cell. Mature, differentiated cells are usually sluggish when it comes to their ability to reproduce as well. Nevertheless, in certain of the body's tissues or compartments, very active natural regeneration occurs. Skin or blood cells, for instance, only live for a relatively short time, necessitating the continuous replacement of dead cells. A special kind of cell is responsible for this, the skin or blood stem cells (in general: adult or body stem cells). Adult stem cells are immature cells whose development has, as it were, got stuck half way. These cells have two particular properties that distinguish them from all other cells: they can develop into mature, fully functional cells and reproduce themselves almost indefinitely. If identified and extracted, these cells can be successfully used for transplantation into patients. It is routine practice to treat leukaemia patients with blood stem cells derived from bone marrow and burns victims with tissue preparations obtained from skin stem cells. However, it remains impossible to adopt this approach for tissue or organs with no natural potential for self-renewal, or whose natural sources of regeneration have not yet been identified. Stem cells from the skin or blood have limited development potential. A blood cell or neuron cannot develop from a skin stem cell.

As an alternative to such adult stem cells, it is possible to use even less mature, so-called pluripotent cells that are found at the very start of an organism's development. In 1981, Martin Evans and Matthew Kaufmann discovered embryonic stem cells in the mouse by extracting cells from very early-stage mouse embryos and getting them to reproduce indefinitely in Petri dishes. Around 20 years later, James Thomson published the derivation of human embryonic stem cells. Embryonic stem cells are not limited in their development – they have the potential to turn into any cell type found in the body.

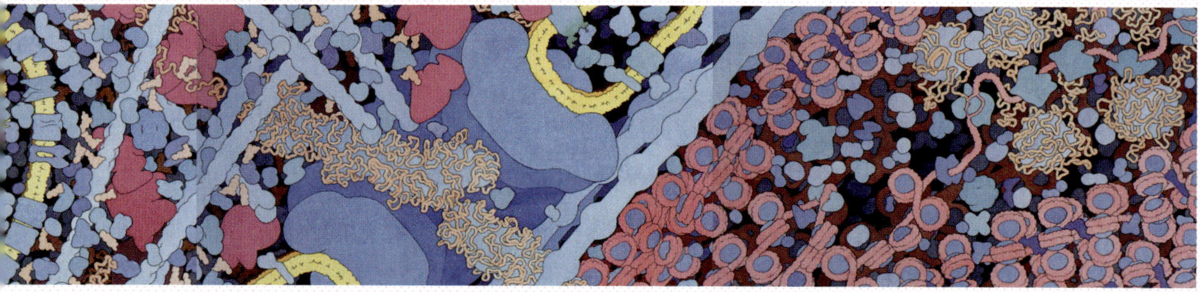

Development into the desired cell type can be controlled by adding hormones, growth factors and other extra-cellular signals. Animal experiments have already shown that such laboratory-bred cells can be transplanted, and that these transplanted cells can at least partially fulfil the intended function. Despite this initial, encouraging success, a therapeutic use for cell design from stem cells is still a long way off and involves a host of unpredictable factors. A whole range of cell types, for instance, are impossible or extremely difficult to access for breeding in a petri dish. And the problem of how individually cultivated cells that are then integrated via transplantation can functionally reconstitute the mature cell aggregate of a complete organ will require intense research. Besides the biological limitations, a serious ethical conflict is surfacing: the potential alleviation of suffering by means of stem cell therapy versus the use of embryos to produce the embryonic stem cells. Although the embryos needed for the preparation of embryonic stem cells are so-called surplus embryos resulting from in vitro fertilisation procedures, this does nothing to alter the fact that, in principle, they are viable cell formations that are destroyed in the course of preparation. In order to solve this ethical conflict, current research approaches are seeking alternative ways of producing pluripotent stem cells. There are some initial indications that the dedifferentiation or reprogramming of already developed cells into immature stem sells could be a feasible course of action. Another *generatio spontanea* phenomenon, then, similar to the widespread opinion that prevailed well into the 19th century? No – for these experiments also show that nothing is possible without cells – there has to be at least one cell at the beginning. And it seems highly doubtful whether humans will ever be in a position to change that. Virchow would probably be both delighted and inspired by the modern development of cell biology. One of the aims he formulated was to achieve "freedom, together with its daughters education and prosperity". He worked towards this goal not only in his function as a doctor but also as a politician of the Progress Party. He pleaded for a liberal society and social medicine based on natural-scientific enlightenment – a motto that remains highly pertinent in our own times.

PETER SLOTERDIJK

THE THING ABOUT POWER

THE THOUSAND YEAR EMPIRE OF COMPETENCE: ON THE CONCEPT OF SELF IN THE MODERN WORLD

Almost everyone who has thought deeply about the essence of modernity joins in a more or less explicit consensus that the age in which we live constitutes an enormous experiment by the primary nations capable of deploying technologies on the subject of the unlimited increase in power and the incessant intensification of life. In this world experiment modern individuals – progressing from generation to generation – act as the bearers of a specific will to power. It is, however, by no means true that with men of the likes of a Columbus or a Descartes, a Cosimo or a Bacon, overnight as it were a new finished species of theoretical and practical powermongers took the stage, who thereupon in line with a violent master plan imposed and established a premeditated empire of unlimited ability. The afore-mentioned names stand as the personal symbols of a transpersonal upheaval in which the epoch-making experiment whipped up the progressing increases in power and ability. It is not single individuals who are the initiators and first causes of the great Modern emergence into the ontological state of new competences, but it is an autonomous spiral of enhanced competences which cuts across all active subjects that starts to turn with the creative spirits of early Modernism by recruiting the inventive powers and initiative of such men. The big names from the early days of the European history of empowerment are, as it were, the names of experimental apostles – they designate individuals who can be considered the first-appointed of a new global European mission; they were the bearers of an original apostolate of the power of knowledge that reproduced with the irresistibility of a victorious religion, generating new vocations and

Since the dawn of the modern age a power spiral with a dynamic of its own has developed in European industrial nations and it can no longer be reversed. Ever since Habsburg Emperor Karl V coined the slogan *plus ultra* it has been the decisive European catchword of the modern age. Meanwhile a new middle class has evolved that uses the medium of design to achieve ever higher levels of Modernity. Sloterdijk cites three sources for the emergence of design: ritual, the world of consumer goods and applied art. The trend in competence is for man to be increasingly better at doing increasingly fewer things. To prevent him from feeling depressed about this society upgrades his individual performance. Which is where design comes in: it serves to simulate mastery. "In complex devices design creates that façade of symbols and points of contact on which the user can have his game connected up without

inspiring repeatedly new successions. The power and competence extension spiral of European Modernity can be described as a kind of reproduction game in which the respectively new generations (starting from the level of ability of the older one) inserted its unique chapter into the epochal novel of enhancement. Prior to the drama, all actors are equals, and prior to the call to power at the beginning of Modernity the differences of the estates disappear; emperor and citizen are both media of the autonomous spiral of power – like the engineers, the nobles hearken the call of the new horizons of ability. *Plus ultra* was the motto of Habsburg Emperor Karl V, and the Spanish fleets crossed the world's seas of the early 16th century under the same sign, of which one could justifiably say it was the decisive European word of Modernity. Only he who takes part in this ever-forwards in terms of his inner *modus operandi* is in the precise meaning of the term a modern European. Only after the motif of enhancement has become embedded in the pilot individuals of the Modern spiral does subjectivity of the modern type actually become possible and real. In the will to move ever-forward, compulsion and spontaneity become inextricably mixed such that it is no longer possible to say whether the spiral draws its dynamic more from the wish to be able or from the need to be able. All that is certain is that its oscillating mass eventually becomes ever more subjective, i.e., infused with ability, and thus the wish to have the ability becomes the ability to have the wish and the having to be able becomes the ability to have to. In other words, each time the element of competence leads the way and the individual progressively emerges as a subject who founds his action, his knowledge, his desire and his wishes on a subjective capital base of competences. This is all just another way of saying that modern individuals are discreet media of power; modernized power is more than a sluggish treasure; it rests not only on a stock of simple power but develops qua power with an intensification vector, as the power to empowerment, if you will. Anyone who is able to do something in the modern

feeling any real shame at his evident incompetence to understand the interior workings." Symbolically speaking, design is parallel action. The designer repeatedly questions the form and function of things. We no longer have static goods, only "improvements", argues Sloterdijk. The designer creates "outperforming and enhanced items". As such, design is not merely one of many success factors but is crucial to modernized success. Design serves to lend us a certain style. The downside of this illusionism is an increasing soberness. It would seem as if man were moving towards flat processes. In such a scenario the emotional level would be replaced by operator, memory and image space. Sloterdijk concludes from what was said earlier that design will get entangled in what he terms a "psycho-political battle between hope and despair". And in such a situation it would seem impossible to be both fellow combatant and

sense can do it such that this increase in ability (along with a wish to be able to do more) is intended and enacted a priori.

I wish to use these preliminary indications on the dynamic form given power and competence to define the character of the contemporary age. A language game widespread in the Western world suggests we should opine that we live in a postmodern world. This is to be understood either as an epigonal position vis-à-vis a heroic and avant-garde modernity, above all in the arts, or as a now sober position vis-à-vis exalted notions of planning history and mastering nature. If Modernity was a compound of ingeniousness and constructivism, then postmodernism is one of mediocrity and chaos management. I wish by contrast to show that such juxtapositions are untenable in terms of the history of competence. Because the power intensification spiral moves unhindered across both positions and right through them; indeed, one could even be of the opinion that the so-called postmodernism is only another landmark in the empowerment dynamic that has been accumulating for centuries. It is characterized by the social amassing of what were previously avant-garde qualities and the translation of once pathetic creativity into everyday manipulation of materials and signs by the members of a world-encompassing design civilization, i.e., by the transnational new smart middle classes. In this regard, Modernity and postmodernism are linked by an overwhelming continuum. Without doubt, postmodernism is likewise only a phase in the history of Euro-American "*plus ultra*". It has not brought a relaxation in the compulsion to power with it, and instead at best adapted the having-to-be-able to the latest state of technology and introduced a little leeway into the contemporary style of competence. There are no signs of a true change in epoch in the sense of a discontinuation in the escalation of competence. As long as a higher power does not shatter the spiral of intensification of ability, it continues to be taken that one twist further as the kinetic heart piece of Modernity. Even what wishes to resist it seems

observer. Though designers work with optimism at producing innovation, they are desperately treading the self-same paths. Their indecision reflects the mental state of all those holders of competence. Meanwhile, a transformation of man lies ahead that will extend to include the psycho-physiological and neural processes.

to contribute to its forward momentum; anyone combating it drives it onward. What will follow Modernity can thus only be another, higher level of Modernity. As long as the course of the world is left to this intrinsic dynamic of its own, nothing lies before it other than the continuation and enhancement of itself to the point of the unforeseeable and yet essentially always identical – the borderline values of which one nevertheless assumes that they could be glossed over and suspended ever further. Modernity is thus its own Endzeit, and in essence it can be nothing other for itself than the source of its own future to the extent that it remains a turn on the spiral of competence. In this sense it needs to be understood as a dynamic millennium. Whoever leads their lives as genuine participants in the world experiment of Modernity must justify to themselves why they spontaneously participate in this millenary operation. And ever since the European 16th century it has come under what remains an insurmountable imperial motto. Irrespective of whether we see Columbus or Karl V, the engineers of the Renaissance or the utopians of the Baroque as the original apostles of the Modern gospel of competence, with each contemporary expression of our lives we are still in its line of succession. Whether we look backwards for centuries or extrapolate the present into the future – we are initially and usually agents and media for a Thousand Year Empire of Competence.

SIMULATION OF SOVEREIGNTY: ON THE BIRTH OF DESIGN FROM THE SPIRIT OF RITUAL

As far as we know, this troubled empire is not some comfortable place. In the agitated Endzeit which knows of no end, a novel, competition-like meritocratic constitution is impressed on the life of the individual; this compels the individual to develop into a ready-to-adapt bio-machine. As an able person or the owner of competence, the individual subject must become a bearer of an abstract willingness to achieve/perform and factually possess the ability to perform. His social pride and private dignity are based in the awareness of his ability to contribute to a totality of intensifying achievements. The individual in the universe of competence must grasp himself as the relative sovereign in his sphere of action. Precisely in this way, the modern individual finds himself in a trap from which there is no escaping – or at least no direct escape. The trap invariably springs open ready to snare him owing to the fact that the individual proud of his achievements in the age of competence enhancement is but a smaller, constantly relativized and specialized person in the overall vortex of the competence spiral. The modern able person is becoming forever better in an ever more limited field. What is on the one hand the justified basis and reason for his existential pride (the sharp-minded mobilization of intention and ability against an open horizon) becomes, on the other, the basis for his fundamental and inescapable disgrace. The mass of competence in the experimentally mobilized world as a whole grows exponentially in relation to the progress in learning made by the individual bearers of ability. The more competence the individual acquires, the more certainly he becomes a player in the overall game, and it is one in which the radius of his expertise (however great it may be) must appear to be negligible. This paradox of the simultaneous rise and fall in individual competence forms the background against which the system of Modern individualism develops. Individualistic civilization faces the paradox task of ramping up the abilities and standards of the individual to such an extent that the ambitiously spurred competent individual does not become terminally depressed owing to the inevitable discovery of their now gradually visible immeasurable incompetence in everything else. Individualism creates the energized psychosocial climate that both provokes and annuls the sovereignty of the individual. The principle of design finds its place in the system precisely as this embarrassment starts to fully unfold. For seen from the vantage-point of an ecology of competence, design is nothing other than the able processing of that which we are not able to do. It shores up the limits of competence by placing at the individual's disposal those procedures and gestures that enable him to navigate ably in the ocean of his own incompetence. To this extent, we can define design as the simulation of sovereignty: Design is if you are able to after all.

I believe it bears going into this matter in a little greater depth. As one can infer from the above, it is by no means directly connected to the manifest topic at hand. Just as Martin Heidegger passed down a famous verdict when he insisted that the essence of technology was itself nothing technical, so, too, as regards our subject we must elucidate that the essence of design itself has nothing design-like about it. I have just defined design as the ability to not be able, and I would now like to underpin this formula with a few anthropological considerations. The roots of able inability of course reach back well before the modern world of competence; indeed they stretch back through the entire field of human primordial history and early cultures; during the latter homo sapiens drifted as tool makers and myth-tellers in hordes and tribes through a natural reality they had still largely neither mastered technically nor penetrated analytically. For them, a lack of ability – the state of *not being able to do much, and not being able to change much* as regards their environment (at least compared with the radius of power in late culture) was first nature, as it were. Nevertheless, early man was anything but some helpless, fear-filled victim of an overly powerful outside world. On the contrary, he was a vibrant, inventive and highly flexible actor in a game of survival which he played with great success – even if in his wildest dreams he would only have equated the competence horizon of a mediocre modern individual with an existence with divine super-powers. If today his form of life may seem to entail a culture of sheer impotence, then our eyes deceive us. In truth, modern individuals are far more endangered by impotence than was pre-historical man owing to the broad development of the plethora of their abilities. They more frequently risk the experience of failure due to incompetence, and across a broader front. Early man, by contrast, benefited from usually knowing almost all the movements required for personal and social self-preservation, while surviving in a more or less routine manner everything he was not able to do by relying on the protection of rituals. Just assume the deluge hits your roof of leaves, complete with thunder and lightning, then, if the flood even allows you to survive, you will survive it better if you recite a song to the weather gods. It is not important to be able to make the weather yourself – even the modern techniques do not quite get that far – but to have a technique at hand to stay in form during bad weather; it must be within your abilities to do something even when there is absolutely nothing else you can do. Only he who knows what to do when nothing more can be done has sufficiently efficient games for life that help avoid breaking into anomic panic or freezing up such that your soul goes rigid. Skilful non-ability engenders a kind of running-on-empty or a parallel process in which life can continue even in the presence of things that leave you impotently faint. I will use the terms ritual, taken from ethnology and religious studies, to describe these parallel processes. Through rituals pre-historical people sense the existential earth under their feet; rites seem to be the material from which the coherence of the world is forged. Although even pre-historical man could not be quite sure that the sun really goes up because they were awake before and encouraged the sunrise by dancing in a circle; but in this way they were able to cope with the demons of dawn and were able to ritually play their way into the day and preserve their mythical identity as the children of the bright star and the dark earth. As of archaic times, the gap through which impotence, panic, and death enter life is closed by ritual. In this sense, we can speak of the birth of design out of the spirit of ritual. Because even if design in the precise sense of the term is an unmistakably modern phenomenon, and manifests itself more through things than through gestures, it is its gestural substratum, ability despite being unskilled, retaining shape in the midst of the dis-shaping, that is after all prefigured in that ancient story of gestural and symbolic parallel acts that we still call ritual today. Without themselves impacting on events in the autonomous environment directly, rituals ensure the coherence of the lives as led by their practitioners and in this well-meaning sense possess the power to imbue a world that cannot otherwise be mastered with order. *Extra ritum nulla salus*. Above all for the threshold events in life (and we have no power over them), in particular the death of a person close to us, many modern individuals rely on the residues of ritual competence; it enables them to ensure the continuation of their game in life by means of the minimum scheme of continuing the right way, getting over things, parallel to the event they cannot master. Something similar can also be said of births and birthdays, weddings and separations, seasons and jubilees. They are thresholds buttressed by residues of ritual, the transgression of which calls for a minimum of formal fitness. As an elementary rule of the game and a source of social form, ritual provides the necessary gestural repertoire.

From here it is no great route to return to the current shapes of design-based incompetence management. In times of need, even the devil grabs a paint bucket. When in the 1970s a Swissair plane crashed at Athens airport, alongside the obvious attempts to rescue the survivors and care for the bodies of the dead, the airline company notably instructed an airport worker to immediately paint over the white cross on the red background all too visible from afar, resplendent on the tail fin of the shattered plane. One could term that first aid for an unfortunate corporate logo. It shows with some precision what design wants and can deliver in an extreme case: the painting over of the tail fin is proof that something can be done even when there is nothing that can be done.

However, it would be frivolous to develop the design question solely from the incompetence catastrophes of the afore-mentioned kind. The larger part of modern professional life and everyday leisure time consists of a competent handling of conditions and appliances for which one cannot really be competent. For the average user all the technical systems that function on the basis of higher precision mechanics, using combustion or nuclear technology, electricity or electronics, are completely opaque entities. Nevertheless, our life has long since been given its place in handling such technology. The basic machines of today's world, the clocks, automobiles, computers, the machine park of entertainments electronics, the higher tools and the like – all of them stand for the absolute majority of users only for glittering surfaces, whose inner workings can never be seen, unless in an amateurish and destructive way. According to the traditional rhetoric, we could speak here of a book with seven seals, or in a contemporary vein such impenetrably complex blocks in users' lifeworlds are black boxes. In the wake of the technological revolution, individuals' lifeworlds are filled with such appliances that empower the user to conduct telepathic operations analogous to magic – such as tele-phoning, tele-visioning, tele-controlling and tele-texting – all of them achievements that rely on internal processes within the apparatuses that have turned their back on the user. Design inevitably comes into play where the black box has to offer the user a contact surface to be of use to the user despite the internal hermeticism. Design creates an open outside for the dark enigmatic boxes. These user interfaces are faces for the boxes, as it were, or, to be more precise, the make-up for the machines; they simulate a kind of affinity between humans and boxes and imbue the user with appetites, a wish to touch, handy sensations and a sense of initiative. The more intangible and transcendental the inner workings of the box, the more appealingly the box's face must smile to the client's natural face and signal: You and me, we can work together; I use my PVC physiognomy to express my sincere sympathy for you and my willingness to serve. By means of design the conviction can be forged that a man and his electric shaver are team mates, as are the woman and her washing machine. Design creates a façade for complex appliances that consists of symbols and points of contact to which the user can, without any apparent loss of face given his evident incompetence as to the inner working, hook up his own game of life. From the user perspective, a lack of knowledge must become power. I fax, therefore I am. The universe of product design revolves mainly around the sensitive subject of servicing the need for competence of users who are structurally incompetent. From this angle, a client is always an idiot who wishes to purchase mastery. And in a strategic alliance with the manufacturers and experts on the inner workings of the black box, the designer is always ready to generate or accomplish new twists and turns on the mastery market. As a user of opaque technology, the modern client is a charlatan who has descended to the level of the everyday – a illuminist with a switch and a dimmer, a telepathy artist with a fax machine, a kinetic trickster at the steering wheel of a car, a levitation master in an airliner. And to the extent that all these dark technical objects would not exist as they are without the contribution of designers, we can term the profession of designer as being that of an outfitter of charlatans – the designer provides everyday charlatans like you and me and everyone else with the accessories for our ongoing simulation of mastery. In the vernacular this tends to get termed support in enhancing life. This service takes its cue from models and related qualities from a sphere that seems quite far removed from if not diametrically opposed to the domain of technology, namely rhetoric, with the teachers in rhetoric and grammar in Ancient Greece and the teachers of dance and good manners in the aristocratic age. Both sets of teachers offered training in linguistic and physical stance that spared individuals crashing into a loss of words or spine even in situations where the ground was torn out from under their feet. If no word fits any longer, then it is still fitting to utter one; where no lip is called for one must still maintain a stiff upper lip. Design repeats this equipping with the means of mastery against the horizon of technological civilization; it provides the technical tools for power for humans who attempt to avoid being exposed as impotent competence marionettes caught on the immense power enhancement spiral today. Whether this attempt will succeed is something the humanist and technoid sides to cultural critique are still arguing over.

COMPETENCE AS THE ABILITY TO REVISE: DESIGN UNDER THE CATEGORICAL IMPERATIVE OF THE COMMODITY WORLD

Having talked about the birth of design from the spirit of ritual, we must now talk of a second specifically modern source of design civilization. The modern world as an experimental culture is in operational terms the practical realization of the conviction that things are not beings or creatures but functions or materialized actions. Were things beings by their own right and origin, i.e., things by God's grace, as it were, then the attempt to get a handle on them would be latently or overtly blasphemous; every design, to the extent that design signifies the redrawing of things, would then be a revolt against their innate or congenital essence. If things are, however, the medium of functions, then they are not protected and sanctified by some seal of their origin, and are intrinsically open to their constant improvement and re-creation. In this sense, design as a stance and profession is rooted in the elementary revisionism of pragmatic Modernity; revisionism is, however, amelioration, making anew means making better. Design is, in other words, the form in which functionalism is realized and perfected – he who conducts design professes to be a practicing functionalist, he is a perpetrator of the verb "to function", an apostle of the belief, taken out into the entire world, that function has priority over structure and essence. Let us take a step back from such self-evident notions and enquire into the meaning of these all-too-obviously understandable expressions, then we will find ourselves in a terrain where the linkage between thing and function, or between function and its thing becomes visible in a decidedly dubious light. In his notoriously somber lecture on the "thing" Martin Heidegger described the questions raised here by taking the example of a jug. The function of the jug (and we need not waste many words on this) is revealed by its suitability for the task of holding water or wine in its hollow interior and being available to pour the liquid out. This is the reason why its appearance necessarily unifies three features: hollow interior, handle and spout. The thing's function would thus simply be the service or benefit it offers. In terms of this example, things are generally speaking useful objects that are at hand. But as objects serving us, things are likewise discreetly sovereign donors – donor-beings in the hands of mortal beings, as it were. This can be seen especially clearly with the example of the jug. The jug is qua office there to pour things out, meaning that it clearly shows how this thing by serving also contributes/gives something. Let us concede the point that Heidegger rightly saw no reason to shy away from the statement that the essence of the jug is to contribute. Seen in this light, it is a small step to the main thing-ontological proposition that the essence of the thing is to be the given giver. This surprising theorem results from a dual understanding of the thing – one that places the functional service the thing gives at the beginning of the deliberation, and then moves on to the human as master and user, and the other, that presumes the character of the thing as given giver and construes humans as the recipients of the gifts of things. In terms of tone and logic, the second opinion is of course at home in a pre-modern interpretation of the world and existence, because it draws to the individual's attention that he should be grateful to the giving object rather than relying on his will to enhance his competences. It constitutes the position of anti-design per se. Anyone who enacts it true to its meaning would not be a competent-incompetent user of things offering power in search of mastery, but a meditator and a recipient, pious in his reverence of things, of gifts in the guise of tools, fabric, and food. *Cum grano salis* this would correspond to a Catholic philosophy of crafts and farming; according to it, any use of tools or apparatuses should rightly always start with a prayer to the things, just as each meal starts with a prayer.

No designer has ever arisen in this way. Designers may consider themselves all sorts of things, but they are certainly not God's assistants or workers in the vineyards of being. A designer can never see himself solely as a curator of the extant. All design arises from an anti-reverence; it begins with the decision to re-pose the question as to the form and function of things. Masterful is he who decides in questions of form on a state of emergency. And design is a permanent state of emergency as regards formal matters relating to things – it declares an end to modesty vis-à-vis the traditional views of things and manifests a will to reformulate all things in the spirit of a radicalized question as to the function and its master and user. Innate in all functionalism is a spark of the Luddite. While one should not think of the price when a gift is concerned, the designer thing is from the outset open as regards questions of price and revisions: Instead of taking the thing the way it presents itself, design puts the function at the beginning and turns the thing into a variable fulfillment of the function. Design is possible because and to the extent that the hypothesis is right that everything comes at a price.

One should, of course, tell the story of the ascendancy of design to the almost unlimited power over the re-formulation of things in an economic tone, too. For what is termed a thing here in ontological jargon, is unequivocally called a commodity in economic terms. A thing that bears a value is a good. If a thing bearing a value is brought to market in order to compete with things of a similar orientation, then the good, if it is willing to be successful and capable of success, will compete with those of its kind to see who becomes the comparably better good, i.e., it will become an article of betterment. This may seem initially to be a play on words – but at second glance it is the valid concept for the value-added object that has been rendered dynamic. The article enhanced to its betterment is a thing bearing value and seeking success and in terms of its dynamic mode of being in the market is per se something that seeks comparison in order to decide it in its favor. One could say that it obeys the categorical comparative: Always present your appearance on the market of goods such that the reason for your being can forever be understood as the expression and stimulus of the striving for betterment! Now because precisely design goods are generated per se as embodiments of the claim for preferentiality compared with other competing goods, they are as it were the real extant comparatives of the things. In the modernized commodity world there are no static articles (to consider the ideal type) in light of the trend of market occurrences, and only betterments – no stale qualities, only articles seeking to outbid and enhance one another. The revisionist thing-concept in design is formulated precisely at the interface between experiment and competition, between improvement in function and improvement in utilization. Now, in addition to these two improvements there is a third, if one considers that a design thing seldom if ever arises on its own. Each individual design object benefits from the proximity to others like it – it absorbs an atmospheric value-added from them that stems from the family affinity with related optimized, stylized, newly-conceived, advanced and exaggerated products, i.e., from betterment. The critical theory of the assortment focuses on betterment groups. Whether it be grasped as part of an ensemble or as an individual item, after rejuvenation through design, the thing is always a comparative object – it is the successor of a replaced or out-bid thing, the result of a history of optimization that is open-ended. If the designer as *homo aesteticus* and *psychologicus*, is, as stated, a supplier of simulations of mastery, then as the *homo oeconomicus* he is the equipper of goods en route to betterment, the man of the unconditional comparative – the aid worker for aspiring things. One could term the designer as the all-rounder for thing revisions. In this regard, he functions as the kit-man for the power struggles between owners for the variable capital that circulates in the form of articles of betterment. And to the degree that the current world market actually rewards betterment, design will not only become one factor among others for success, but indeed a basic element and the nutritional solution for modernized success, i.e., for success rendered more intelligent.

ART APPLIED TO THE EGO

There is a third source after ritual and capital from which design in the current power domain derives meanings. The keyword here is applied art. I do not wish to embark here on an excursion into the swamp lands of theories of modern art. And I shall likewise ignore the Marxist classic of commodity aesthetics and the liberal hit of consumer aesthetics. In other words, I am not speaking about "the role of the aesthetic in the apparent solution of the fundamental contradictions of capitalist society" – we are no longer in the 1970s. I presume that my readers know how designers as the mask-makers of commodities participate in generating added-value through appearance. And that they likewise know that semblances of betterment, the pretense of differences in quality, the creation of the illusion of choice for clients have for a long time now constituted problematic domains of an aberrant form of design as applied art. This is simply presumed and I shall apply it here without further commentary. In a culture of identity, difference necessarily becomes a scarce resource.

As regards applied art, it is, as everyone knows, not only a meeting point for encounters between beauty and techniques, but also the epitome of procedures for regenerating the semblance of beautiful life. To this extent, applied art has privileged access to the dream factories without whose contribution the complex psycho-political machinery of modern mass societies could not be kept in motion. Modernity is, as even the ideology critics now seem to recognize, but another name for the embarrassment of having to strike a balance between the reduction and the establishment of illusions. Design as applied art is therefore always also a regulator in the subjective ecology of an individualistic civilization; it conditions the air of nervous mass societies and plays a role in fine-tuning the illusion and élan systems. It motivates the players in the competition lotteries of the achievement and experience society and tones up their muscle by doling out the premium of mastery as well as the means of its simulation as broadly as possible. Everyone is meant to have access to such feelings of being a winner – such is the rule of inclusive games. As long as it is democratic concepts that hold sway in advanced illusion design, technological progress will always present itself as a lottery for many, if not all. Thus, France's SNCF has taken its high-speed-train strategy out to the people with the slogan: *Le progrès ne vaut que s'il est partagé par tous*. However, a hard-nosed sobriety is emerging as the obverse of this generous illusionism. Its symbols ghost their way through all the media, and the trend press has for some time been broadcasting on the same wave length. Applied art – with a new lack of illusion combined in exclusive games – leads to the modernization of egoism, and this result of the new mass-based *self-designing* served to breathe cold air into the hothouse of postmodern illusion in the West prior to 2000. Design as art applied to the ego generates a highly topical bundle of competence for its smart bearers, consisting of speed, information, irony, taste and second-level inconsiderateness. The previous avant-garde idea of transforming the life of each individual person into an artwork has now, after a lag of just under three generations, reached the grass roots. What many call life-style is the breakthrough of design at the level of self-stylizations and personal biographies. The individual now tries to grasp the competence of being able to realize himself as a compromise between artwork and machine – e.g., following the example of Andy Warhol, who has long since been treated world-wide as the patriarch of design-supported Neo-Individualism. Later generations have learned from him that mastery is an effect resulting from investing energy in flat processes. And to the extent that the individual in the design age wishes himself to become the operator of flat processes executed on his own body, we can look forward to seeing the advent of a new psychosocial era, perhaps even an anthropological quantum leap. In the following era, if we are not wrong in discerning the trend, we will inevitably see a change in the gestalt of the traditional human imago, right through to a new face given to psycho-physiological and neuronal processes. It would seem that a type of *homo semioticus* is about to replace the high-culture *homo psychologicus*; the manifest bearers of this development have already reached the age of maturity, our children, our mutants. In their case, the classic "deep" triad of psyche, memory, and inner world has been replaced by the new flat trio of operator, memory, image space. The "soul in the technological age" could become something like the living cursor in turbulent experiential spaces – a cursor seeking its curriculum, a runner seeking a track that he can call his "own".

BATTLE OF THE TITANS

Irrespective of what judgment one may pass on such assumed trends, design is at any rate embroiled in a kind of psychopolitical battle of the Titans in all three of its main strands, a battle in which the powers of hope and those of despair wrestle with each other like two world powers or atmospheric totalities. To this extent, the collapse of Communism by no means spelt the end of what has been termed the bipolar order. One can at most say that the superfluous battle of the Titans, the East-West bipolarity, has disappeared to make space for the necessary battle of the Titans – that fateful battle throughout humanity between the confidence (and everything that gives it cause) and despair (and everything that fosters and intensifies it). It is a battle for the life basis of a humanity that had in the course of its modernization to learn to see its relationships with sober eyes. All contemporary labor and art is to be situated in this bipolarity; the drivers of life for current and future generations come awake in the battle of motifs that give cause or hope or drive us to despair – or through it dissolve. This psychodynamic endgame of species intelligence has largely not been seen by the media, although they all have long since been screaming in the fog like combatants; it has hardly been grasped by the political castes although they themselves long since operate in more or less confused maneuvers on the battle field. No institute for strategic studies has ever scrutinized the course of the ultimate psychopolitical drama, let alone dared propose how it will end. And there is also a strange spell cast over intelligence, preventing it from duly perceiving its role as guardian of the things of the world. Evidently people today still find it difficult to be both combatants and observers in the midst of a battle of the Titans. The rule would seem to be that he who sees does not fight, and he who fights does not see. However, seeing combat and combative seeing would be in order – above all because hardly any one still knows or can know for which side of the battle he or she was actually hired. That is the backdrop to the crisis of visions perceived everywhere today.

Vision as such is blurred by the confused battle of the Titans. The one person's reason for hope is the other's cause of despair; the one person's despair is the other person's hope. And even this last bipolarity has its junk spaces and no-man's lands; the double agents move lost between the fronts, and the scenarios of a deep world war are hidden behind the clouds of ambiguity. Where ambivalence prevails, design is never far away. Designers are also the mask-makers of groundless confidence and creators of the means of simulation for deceptive hopes and false ways-out. They are the core troop of double agents of the titanomachy, for they work with confidence to create the new and the feasible for the future and with the despair of panicking self-survival blindly go tearing down the same paths ever anew. They treat both sides as right and arm both with symbols and appliances. As members of the unresolved class *par excellence* designers are also the suppliers of toys for the last men and inventors of tools that are meant to prove their worth in future workshops. But the designers' irresolution is not mere caprice or a private weakness; it reflects the mental state of all bearers of competence in the current moment in the world's history. It shows that we at present do not know what competence we need were we to give no further grounds for despair through things.

KOERT VAN MENSVOORT

------- EXPLORING NEXT NATURE -------
NATURE CHANGES ALONG WITH US

INTRODUCTION

Almost every one of us loves nature. Still, what is it? For some of us nature may embody harmony, soundness, and peace. For others, nature is rather wild, brutal, and unpredictable. Nature is untouched by man – it is still unspoiled. But paradoxically, man himself has evolved out of that very same nature. "Nature likes to hide itself", proclaimed the ancient Greek philosopher Heraclitus,[1] as early as the 5th century BC. If there is one spot that deserves to be labelled as "natural", it is the environment in which humankind evolved, many, many years ago. This environment has been the basis of our perception of reality and, indeed, of all information we gather. Our human constitution and our senses have been fully adjusted to it. Today, this environment is completely governed by us. It has lost all originality. How natural is it to have a nine-to-five job, and to go to the office with a suit and a tie? The roofs over our heads, the chairs we sit on, even the trees in the forest – they are all what we want them to be. Just take a look around, and try and find the most natural object present in the space you are in right now. Most likely that will be you.

For animals, which do not shape their own world as radically as man, nature will always be nature. Man, on the other hand, started to bend things to his will. By covering our natural environment with a layer of language, technology and economy, we have modified our living environment. Well-known milestones are the invention of fire, the introduction of agriculture and stockbreeding, the utilization of mechanical production, and the rise of information technology resulting in today's technological culture.[2] This

In this paper, we explore and redefine our notion of nature. We will argue that our current common view on "nature" needs reconsideration. The notions of nature and culture seem to be trading places. Nature, in the sense of trees, plants, animals, atoms, or climate, is getting increasingly controlled and governed by man. It has turned into a cultural category. At the same time, products of culture, which we used to be in control of, tend to outgrow us and become autonomous. The "natural powers" seem to shift to another field. Nature changes along with us. We propose the term "next nature" for this culturally emerged nature. Next nature is real nature, no representation, or a simulation of some long lost phenomenon.

cultural layer is well-functioning and has become our "reality". In a way it is a pity that, at the time, there was no Nobel Prize to honour that early ancestor who came up with the idea to make a coat out of a bearskin, for that most ingenious animal would certainly have deserved it. For ten thousands of years, human development was confined to the snail's pace of genetic evolution and we took no advantage from the numerous things we learned throughout our lives. Apart from some rare mutations, the genes we give to our children are identical to the ones we get from our forefathers. Cultural evolution, however, is much faster. The invention of the wheel, television, the nuclear bomb, the frying pan... they are all crucial things that no person could possibly invent during his own life. Some argue the human race would be better of if we got rid of these inventions.[3] But once they have become part of our daily lives, it seems impossible to imagine life without them. Continuing on Charles Darwin's genetic evolution,[4] zoologist Dawkins concluded that replication also happens in cultural evolution, yet in a different sense. He coined the term meme[5] as the basic unit of cultural evolution.

OUR ENVIRONMENT BECOMES THE INTERFACE, AGAIN

Why do we experience the sound of crickets or the sound of the sea as soothing, as opposed to, for instance, the sound of an engine? Normally, we do not associate technology with peace. On the contrary, especially information technology is often quite the opposite of peace and naturalness. Dial tones, e-mails, websites, television and radio – they overload us with all kinds of data. But does technology only give us stress and disquiet, or can it also be the other way around? In fact, there is also plenty of technology that gives us relaxation and convenience. A pair of sneakers is not necessarily less sophisticated than a desktop.[6] Originally, computers were developed in order to make calculations for business or military purposes. At first, there was only one supercomputer available for the whole company; nowadays, each employee has his own desktop. Only recently, computers have entered our personal living environment. Soon, each carton of milk will contain one. United Nations research predicts the use of robots around the home to mow lawns, vacuum floors and manage other chores is set to surge sevenfold in the next three years.[7] Domestic robots will become familiar companions in our home environment. And when computers are present everywhere, we cannot possibly operate them all ourselves. Therefore, they will start operating themselves, as well as one another. The radio alarm will tell the electric kettle and the toaster that you are awake and about to have your breakfast. The fridge has already ordered your milk. The morning news is toasted on your bread. As soon as you have left the house, the vacuum robot will start cleaning. And when you return at night and stick your key in the front door, the oven will know that the pizza needs to be baked. At your service, your future house will contain an entire ecological system.[8] The most profound technologies are those that disappear. They weave themselves into the fabric of everyday life until they are indistinguishable from it.[9] After all, neither do we experience a warm winter coat nor a fine writing pen as technology. In the early days, machines were coarse, heavy and rusty, but nowadays they have become more and more transparent and they are as bright as sunshine. Our environment becomes the interface, again.

THE INFINITE FORCES OF NATURE

The relation between man and nature can be summarized as conquering and being conquered. Ever since the English philosopher Francis Bacon,[10] we seem to believe that knowledge is power. The principle of progress has never been so popular and generally accepted.[11] It was Bacon's conviction that man could acquire control of nature by means of scientific discoveries and inventions. Nature was to be tortured in order to make it reveal it's secrets. If birds can fly, why should'nt man be able to? For years and years, man has tried to acquire from nature what he himself does not have. And for what purpose? To be able to survive without nature? Just imagine a life without it... would that be possible? I do not think so.

While battling the forces of nature, man has become more and more independent of physical conditions. At the same time, however, he has become more and more dependent on technical means, of other people and of his own self. Just think of the various forms of dependence that come along with driving a car. There have to be highways, for which road tax has to be paid. Petrol supply has to be arranged. Once on the road, you will have to concentrate, otherwise you might end up in a car crash. You will have to show consideration for other road users, and you simply need to have your driver's licence. And all this is required just to move your body from A to B and save a little time. Physical independence is achieved at the expense of social and mental dependence. Highly precise and productive machines often require highly precise and productive humans to operate them.[12] The things we design often end up designing us.

Koert van Mensvoort has been consistently addressing the topic of "Next Nature" (www.nextnature.net) together with designers Mieke Gerritzen and Karl Grandin. The three also teamed up for the "Visual Power Show" (www.visual-power.com) shown in the run-up to ENTRY2006. Here you see Gerritzen/Grandin image of a "Logo Jungle".

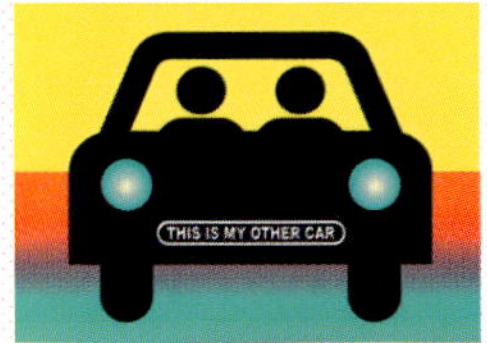

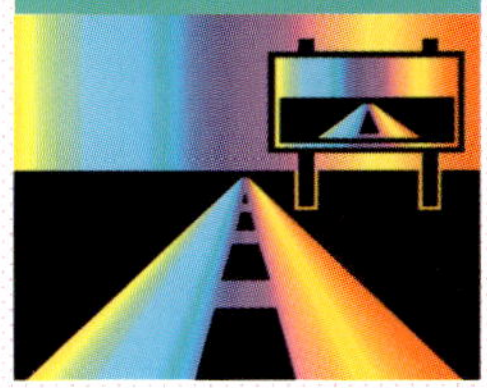

Mieke Gerritzen and Karl Grandin "This is my other car" (above), "The Environment is the interface" (below) and "Next nature pear" (below right).

› Jack van Wijk "Database landscape"

Another example: A fridge is a most convenient device that helps us to keep food fresh. CFCs are stable industrial gasses, which are included in cooling substances, refrigerators, synthetic foams and various spray cans. CFCs have done so much damage to the ozone layer that the latter has become disturbingly thin. The hole in the ozone layer has been discovered in 1985. At first, when British scientists at the Halley Bay base in Antarctica examined the results of their measurements, they were convinced that something was wrong with their instruments. But after these had been replaced, the conclusions continued to be as alarming as before. When their findings were published in *Nature* magazine,[13] it became clear that something urgent was to be done. The ozone layer protects man and animals from the harmful ultraviolet radiation of the sun. If this protective layer is dismantled, this will have dramatic consequences for us all. Prolonged, unprotected exposure to the sun causes skin cancer. The inventor of the CFC refrigerator had probably never even heard of the ozone layer.

Despite our attempts and experiments, life cannot formed that easily. Every time nature seems to be conquered, it stands out as a tough opponent in another field. Hence, the battle between man and nature is infinite. Nature likes to hide itself. And it is not just man who evolves, but, in the process of evolution, nature comes along. My point of view is that perhaps it is better not to see nature as a static entity, but rather as a dynamic power that is beyond our reach by definition – a power beyond our reach, but with which we will have to deal constantly. To put it differently, maybe we should see nature as the permanent sparring-partner of our development.

THE BIOLOGY OF TECHNOLOGY

From our biology classes we all know the evolutionary scheme: the monkey that starts walking upright, loses its coat and becomes a human being. But there is something

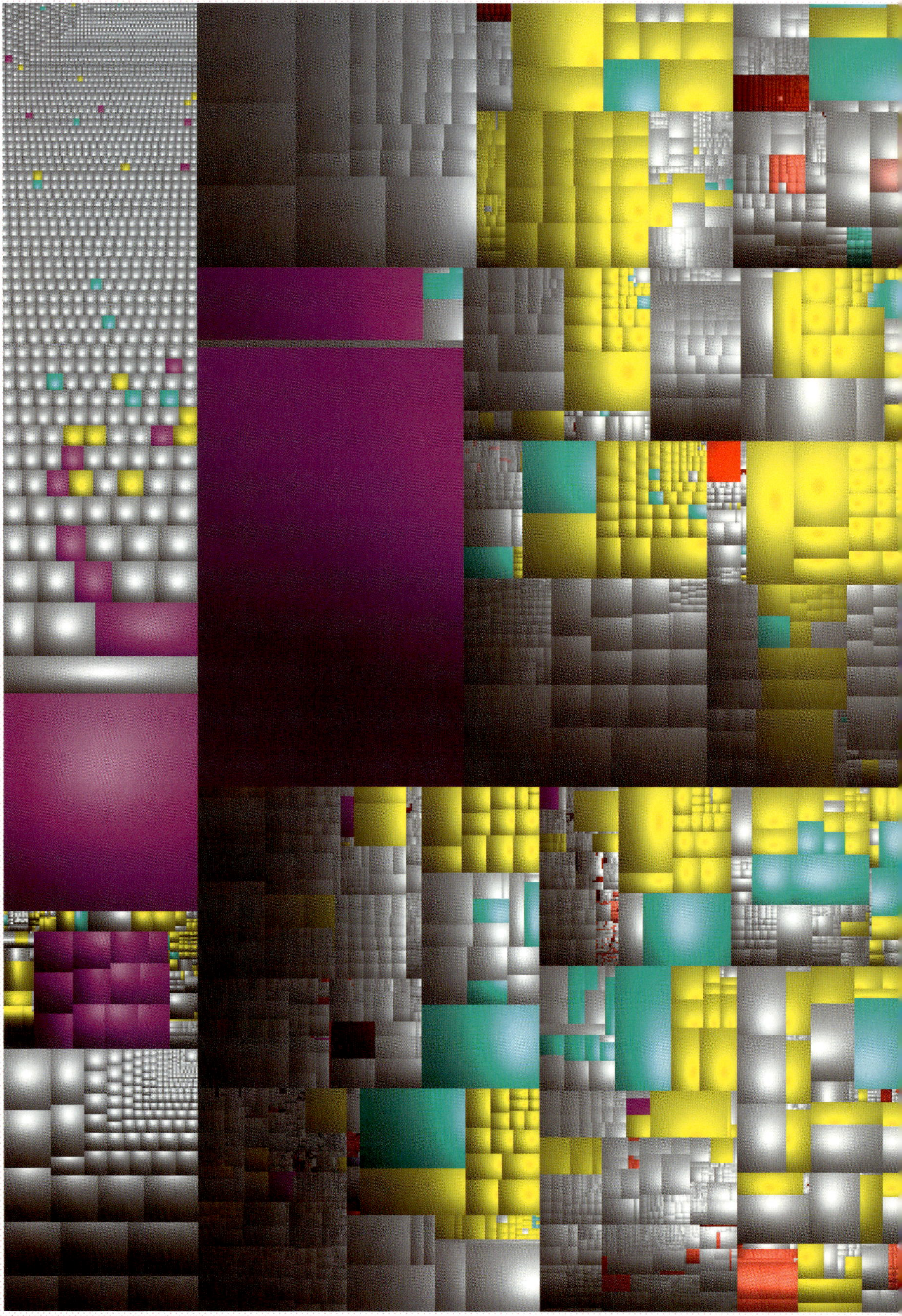

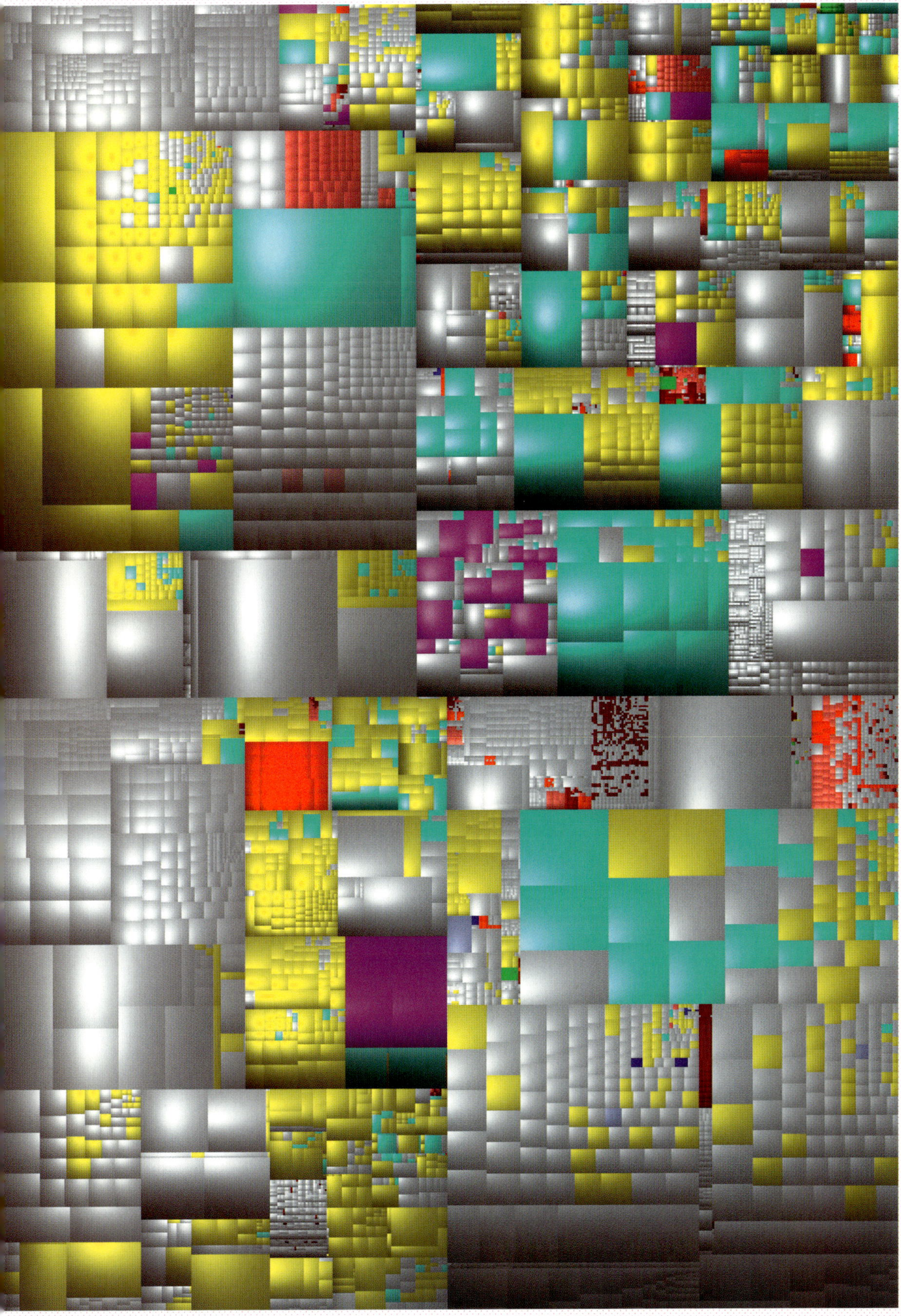

fundamentally wrong about this picture, and no one seems to notice: that human being is naked! Essentially, this is as "unnatural" as a juggling monkey in a clown suit.

What animal is so stupid to enter this world as a naked and crying infant, completely vulnerable, helpless, and an easy prey for any predator? Any newborn lamb or giraffe's baby can walk within a few hours, but it takes man years and years to learn to take care of himself. Therefore, it is no less than a miracle that, despite his physical vulnerability, man has managed to keep dominating the world. Compared to other animals, man is maladjusted, primitive and undeveloped. Still, he has been able to survive, among all those animals that are so much better equipped. Other animals have specific organs, skills and reflexes that enable them to survive in their proper environment. But unlike animals, man has never been placed in an environment for which he is specifically equipped. Actually, the physique of man implies that there is no such thing as a "purely" natural environment: by definition, man is a creature of culture.[14]

To be honest, our picture of nature is fairly naive and definitely up for reconsideration. At various locations in the Netherlands, prehistoric woods are being laid out: our picture of unspoilt nature is being carefully reconstructed in a real-time simulation.[15] The Green Heart (a renowned natural reserve in the Netherlands) is actually an agricultural and industrial area from the 15th century. A city girl who washes her hair with pine shampoo, one day takes a walk with her father in the woods. Suddenly, she says: "Daddy, the woods smell like shampoo".[16] Such a young girl would surely perform better at discriminating corporate logos than tree- or bird species. The more we tend to think and act in terms of symbols, the more we lose our grip on physical reality.[17] Was Mondrian's picture[18] the forerunner of the ultimate natural picture after all?

WILD SYSTEMS

Over the last centuries our world view has become more and more actual – actual in the sense of "in accordance with the truth", but above all in the sense of "explicit and articulated". We think we know how things work: our reality has become a set of facts. Mankind is no longer gazing in wonder at the magic of the moon and stars. We have rid ourselves of the belief that our environment is governed by mysterious powers, by spirits and demons. Max Weber has aptly called this "the demystification of the outer world",[19] alluding to the fact that the pre-modern world was still thoroughly ruled by quirks of fate that left mankind powerless. Fate could have a devastating influence on our lives, but could just as well move our lives into a more favourable direction. After the modernization of the world, man has found himself in a much more rational world, in which everything is schemed or calculated. Bureaucratic structures as well as control models and strategies are being set up to promote a more prosperous world. The rhetoric of decision-making is more and more ruled by rational thinking, and logos seem to dominate our culture. And when we say myth, we often mean that something is plain nonsense. However, the greatest myth of our time may well be the primacy of the logos. Systems seldom function the way they were meant to by their makers. Administrative malpractice as experienced by multinationals such as Enron,[20] Kmart and WorldCom[21] took their respective shareholders by surprise and proved to be genuine disasters. Ironically, we are able to protect ourselves from floods, earthquakes and tornados, but we are powerless to do anything against this sort of bookkeeping scandals. Quite recently, the oil reserves of Shell appeared to be much smaller than was forecasted. But that was not the fault of the oil reserves; quite the contrary, it was the fault of the Shell management.[22] Systems are the products of culture; still, they can outgrow us and become inaccessible, and force us to "re-cultivate" them.

In the future, natural disasters may cease to exist. In fact, they seem to have already done so. After all, at every disaster we start looking for a scapegoat in the form of some bureaucratic institution: Why were the dykes not high enough? Who is responsible for the earthquake safeness of that building? Natural disasters have turned into cultural disasters and organisational failures, and we will certainly have plenty of them in the years to come. Systems can be rude and merciless.

OUR NEXT NATURE

"Some weeks ago I took a walk in the woods and bumped into the biggest rabbit I had ever seen. The rabbit was already nearly white for the winter and we stood eye in eye with each other for a short and pleasant while – two creatures joined together by mutual curiosity. What will it mean to come across a rabbit in the woods after genetically engineered 'rabbits' are widespread? Why would we have any more reverence, or even affection, for such a rabbit than we would have for a Coke bottle?" It is in such terms, in his book *The End of Nature*, that Bill McKibben recounts his mythical encounter with a white rabbit in the woods.[23]

Let us look at it from our own perspective: nature as human experience. The associations that most people have with the notion of nature can be summed up in such terms as infinite, inaccessible, overwhelming in power, primal, wild and fearsome.[24] But where can this kind of nature be found nowadays? In the park on the outskirts of the town? Or on the window sill, where your cat is gently sleeping? Probably not. Our next nature arises from cultural products that have become so complex that the only way we can relate to them is in terms of a man-nature relationship. There may even come a moment that our connection with an industrially manufactured Coke bottle may be richer and more mythical than our relation with a genetically analysed and manipulated white rabbit in the woods.

In point of fact, the notions of nature and culture are trading places. Nature, or whatever we mean by it, is getting more and more governed by man. Nature, in the sense of trees, plants, animals, atoms, or climate, has turned into some sort of cultural category. At the same time, products of culture, which we used to be in control of, tend to outgrow us more and more. Those "natural powers" seem to shift to another field. Our next nature will consist of what used to be cultural, even though this does not alter the fact that this is still a "real" kind of nature. No representation or simulation of some long-lost phenomenon. On the contrary, it is as real as can be, and has all the workings, threads and opportunities of the older, natural phenomena. Wild systems, genetic surprises, calm technology, autonomous machinery and splendidly beautiful black flowers.

We seem to enter a magic garden that may take us by surprise and astonish us, and that may also knock us down or be kindly disposed to us. We will have to make sense of things all over in order to define our place in the world; we will have to invent a new symbolic meaning of things. As newborn children, we wander through our own creation. Greener grass, you get used to it. Surely I am a nature-lover.

ACKNOWLEDGEMENTS
The author wishes to thank Mieke Gerritzen and Berry Eggen who provided helpful comments on previous versions of this document as well as Geert Lovink for his help with the terminology.

1 Heraclitus (540 BC – 480 BC). *On Nature*, fr. 208: "Nature loves to hide". (Heraclitus wrote a philosophical work entitled *On Nature*, which he placed in the temple of Artemis at Ephesus (Diogenes, Lives, 9. 6). The work as a whole has not survived; what remains of it are quotations from it in the works of others.)
2 Schwarz, Michiel; Jansma, Rein, (eds.): *The Technological Culture* (Amsterdam, 1989)
3 Zerzan, John: "Future Primitive", in: Colombia: *Anarchy Magazine*, 1994
4 Darwin, Charles: *On the Origin of Species* (London, 1895)
5 Dawkins, R.: *The Selfish Gene* (New York, 1976)
6 Weiser, M.; Brown, J. S.: "Designing calm technology", in: *PowerGrid Journal*, vol. 1, No. 1, 1996
7 United Nations Economic Commission for Europe: *World Robotics 2004 – Statistics, Market Analysis, Forecasts, Case Studies and Profitability of Robot Investment*, 2004
8 Dix, A.: Managing the *Ecology of Interaction*. Proceedings of Tamodia 2002 – First International Workshop on Task Models and User Interface Design (Bucharest, Romania, 18–19 July 2002), C. Pribeanu, J. Vanderdonckt (eds.) (Bucharest, 2002) pp. 1–9.
9 Weiser, M.: "The computer of the 21st century", in: *Scientific American*, September 1991, pp. 94–100
10 Bacon, F.: *New Atlantis*, Kila: Kessinger, s.a, 1626
11 Condorcet. *On the Future Progress of the Human Mind*. Modern History Sourcebook, 1794 (original French version: Marie Jean Antoine Nicolas Caritat, marquis de Condorcet. *Esquisse d'un tableau historique des progrès de l'esprit humain, Paris*, 1822)
12 Schelsky, Helmut: *Der Mensch in der wissenschaftlichen Zivilisation* (Cologne, 1961)
13 Farman, J. C.; Gardiner, B.G. and Shanklin, J.D.: "Large losses of total ozone in Antarctica reveal seasonal ClO_x/NO_x interaction", in: *Nature*, 315, pp. 207–210, 1985
14 Gehlen, A. Man: His Nature and Place in the World (Columbia, 1988) (original print: *Der Mensch. Seine Stellung in der Welt*, Frankfurt, 1966).

15 Metz, Tracy: *Nieuwe natuur. Reportages over veranderend landschap (New Nature, Reports on a changing landscape)* (Amsterdam, 1998)
16 van Mensvoort, Koert: *Shampoo Generation (The Woods smell of Shampoo)* (Amsterdam, 2004)
17 Cassirer, Ernst: *An Essay on Man* (Yale University Press, 1944)
18 Mondriaan, P., Blok, C.: *Mondriaan in de collectie van het Haags Gemeentemuseum:* catalogus 1968, S.l.: Haags Gemeentemuseum, 1968
19 Weber, Max: *Gesammelte Aufsätze zur Soziologie und Sozialpolitik* (Tuebingen, 1924)
20 Jickling, Mark: *The Enron Collapse: An Overview of Financial Issues.* CRS Report for Congress, order code RS21135, 2002
21 Patsuris, Penelope: The Corporate Scandal Sheet (New York, 2002)
22 Messenger, Terry: *Oil giant Shell's investors shocked.* BBC News UK, 15 July, 2004
23 McKibben, Bill: *The End of Nature* (New York, 1989)
24 Larsen, S. E.: "Is Nature Really Natural?", in: *Landscape Research* 17(3), pp. 116–122, 1992

ELLEN LUPTON

SECOND SKIN: NEW DESIGN ORGANICS

Is it possible for design to speak about the darker sides of human experience? We expect other forms of artistic expression – literature, film, drama, painting, music – to give voice to a full range of emotions. The demeanor of the fine arts is not incessantly happy and humorous; it sometimes shows rage, despair, fear, or anxiety. Designers, however, are trained, in general, to paint a happy face across the surfaces of life, serving as publicists for an endless supply of optimism and utility.

The body has been design's primary point of reference across history, not only because people like to fashion things in their own image but because functional objects come in close contact with our flesh, skin, and bones. Technologies evolved to supplement the inadequacies of the body's natural envelope. Chairs and jackets, cups and spoons, buildings and cars, serve to extend and support the body, compensating for its failures and allowing it to survive in a world that is largely inhospitable to human habitation. The first shelters and the first garments, made from animal skins, protected people against harsh climates, allowing them to cultivate ever-more hostile regions of the planet.

This essay examines skin as substance and metaphor in recent design practice. Skin is a multilayered, multipurpose organ that shifts from thick to thin, tight to loose, wet to dry, across the landscape of the body. Skin responds to heat and cold, pleasure and pain. Lacking definitive boundaries, it flows continuously from the exposed surfaces of the body to its internal cavities. It is a self-repairing, self-replacing material whose exterior is senseless and inert while its inner layers are flush with nerves, glands, and capillaries.

Ellen Lupton focuses on skin as a contemporary design material and metaphor. There are designers today using skin as an example to experiment with complex structures to design innovative surfaces for products and buildings. Artificial skin does not only alter the haptics, function, and meaning of the object. It can also be frightening. Innovative materials react to light, warmth, touch, and mechanical load. Some have digital and mechanical networks that make them flexible. Others are bred in laboratories as living substitutes for skin. "The dense, radiant surfaces of out, contemporary objects are both beautiful and unnerving, divine and grotesque," writes Lupton. She names chairs covered in skin-like material, and latex clothing as an extension of the natural surface of the skin, as two examples. There are also numerous examples in architecture of a "second skin", like flexible, elastic exterior

Some contemporary designers are approaching the surfaces of products and buildings as similarly complex, ambiguous forms. Manufactured skins modulate the meaning, function, and dimensionality of things. New materials react to light, heat, touch, and mechanical stress. Translucency and mutability have replaced transparency and permanence. The outer envelope has detached from the interior volume. Flexible membranes are embedded with digital and mechanical networks. Thin planes of material are folded, warped, or pumped with air to become load-bearing structures.

Skin is both dead and alive. The thin outer layer, the epidermis, consists of strata of cells that migrate toward the surface, where they compact into a layer of dead material. Skin's protective function relies on the inertness of this outer surface. Mark C. Taylor, whose 1997 book *Hiding* is a commentary on the culture of skin, writes, "Death, like life, is not a momentary event but is an ongoing process whose traces line the body. At the point where I make contact with the world, I am always already dead."[1] Skin, hair, and nails are products of the body, continuously sloughed off and renewed. Hair is part of the skin, its cells generated deep within the living dermis and pushed upward into shafts of protein, emerging across the body's landscape as a thicket of dead blades. Skin is connected to our bodies yet also alien, marking the exterior, the end of our selves. It is a screen on which we witness the body's amazing ability to heal itself as well as its irreversible collapse.

In the 1940s and 1950s, organic forms and materials provided designers with a humanist vocabulary that affirmed society's place within the natural world. A technologically enhanced body emerged, however, over the course of the 20th century, achieving a monstrous perfection and an alarming dependency on creature comforts. From birth, the human organism is enmeshed in an infrastructure that controls and delivers food,

walls, semi-transparent glass facades, and translucent building materials. Artificial skin is still being developed. Some are even prophesizing that brand identities will be transplanted into living tissue. Living dermal tissue is already a commercial product. Objects and buildings increasingly resemble natural organisms. The border between nature and technology is evaporating.

water, light, climate, health care, and entertainment. Rarely naked, never alone, the new body requires constant attention and prophylactic shields against disaster.

Like skin, design performs at the frontier of life and death, body and product. Human beings, using objects to survive and conquer, rely on the world of things, merging their own identities with the objects they use. Photographer Elinor Carucci has used her camera to reveal intimate relations between skin and everyday industrial products, from lipstick and pantyhose to zippers, bras, and buttons. Industrial designers Carla Murray and Peter Allen propose more grotesque conjoinings of bodies and consumer goods in their project *Skinthetic* (2001), which predicts the grafting of brand identities into living tissue. Skin is the surface where bodies and products merge.

Industrial skins have assumed a life of their own. It is a life whose pedigree, however, is more alien than human. Our increasing dependence on the artificial bears with it a measure of anxiety, a phenomenon vividly expressed in science fiction. The term "cyborg" refers to a being who is part biological, part mechanical, enhanced by a host of technological devices. Although the term "cyborg" brings to mind monstrous beings, it also describes modern creatures routinely repaired with pacemakers, hearing aids, and prosthetic limbs, and outfitted with cell phones, PDAs, and wristwatches.

The fear of invasion from within the body drives many depictions of warfare with outer space. In the *Alien* film series, the enemy is most frightening when it occupies a human host, incubating inside the body before bursting through the abdomen or chest. A dream sequence in *Aliens* (1986) shows Sigourney Weaver – prone on a hospital bed – watching with horror as a mechanical-looking object protrudes upward through her belly, threatening to erupt in a hideous birth. The clinical setting heightens the shock of the scene, with its threat of physical helplessness and humiliation.

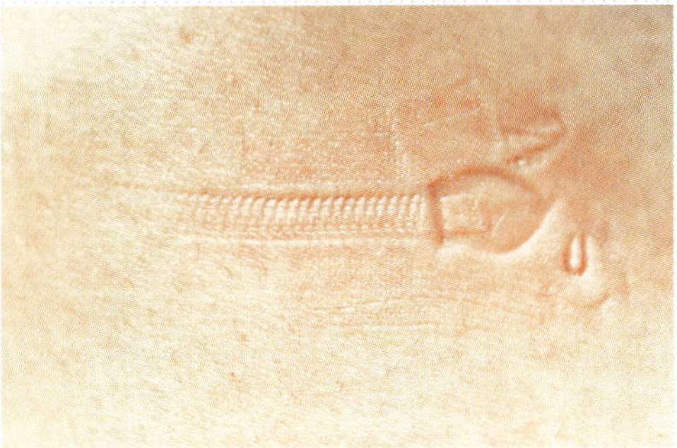

Elinor Carucci: "Zipper Mark", C-print, 1999

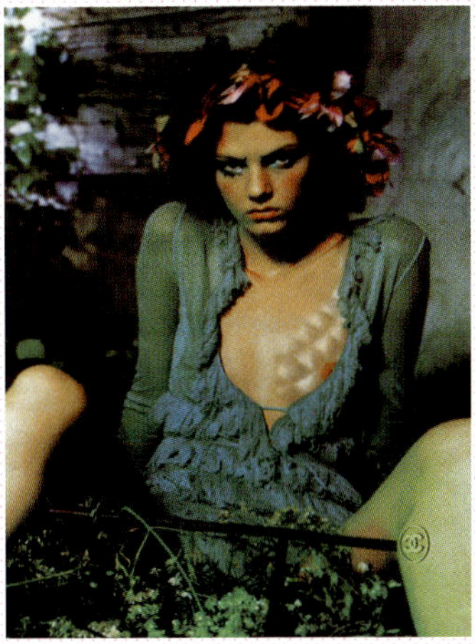

Peter Allen and Carla Ross Allen
(Design, photography and 3D modeling):
"Skinthetic", digital photograph, 2001

Like cinema, design offers imaginative responses to the merging of life and technology, in turn celebrating and recoiling from it. Contemporary objects and spaces are cloaked in surfaces that have been enhanced, simulated, or engineered, surfaces that masquerade as other materials, surfaces where the physical and the virtual, the real and the imagined, collide. Hard surfaces look soft, and soft surfaces look hard. Smooth planes are rippled, bubbled, or scarred with digital imagery; luminescent fabrics, gels, and plywoods glow with preternatural life.

Jurgen Bey's *Kokon* series encloses traditional wooden furnishings inside a tight wrapping of PVC. The humanly scaled limbs of the found objects press through a grossly artificial skin. In "Kokon Double Chair", two chairs are bound, back to back, like lovers held hostage in a dysfunctional embrace. Moorhead & Moorhead's "Rubber Lamp No. 5" has a flexible shade of translucent rubber that peels open to direct the flow of light, evoking the alien pods from *Invasion of the Body Snatchers*, a story whose disturbing picture of the human body as host to alien life forms has been continually retold in cinema (1956, 1978, 2006).

Organic design vocabularies – from the ecstasies of baroque ornament to mid-20th century biomorphism – have always gestured toward the erotic, suggesting the curves and movements of the human body. In contemporary design, eroticism is present yet held at a distance, handled with rubber gloves. The fulfillment of desire and the satisfaction of touch are blunted by protective layers of material. Clothed in latex, vinyl, rubber, or resin, sensual forms are rendered clinical.

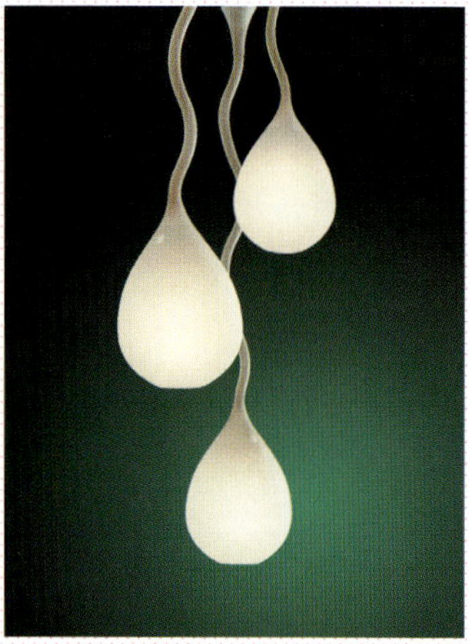

Herzog & de Meuron: "Jingzi" luminaire, silicon and electrical components, manufacturer: Belux

Herzog and de Menron's "Jingzi" lamp resembles a spermatozoa or a large primitive worm. "Jingzi's" electrical components are entirely contained inside a thin-walled silicon tube that also diffuses the bulb's light. The lamp is activated by touching the translucent membrane. Whereas electrical cords are usually inert and at best unobtrusive, the long tail of the "Jingzi" lamp follows the fluid gesture of a sperm cell moving towards its quarry.

TechnoGel is a soft polyurethane material developed in the 1970s for use in wheelchairs and hospital beds, where it supports the body with minimal friction. Werner Aisslinger's "Gel chair" is padded with slabs of TechnoGel, whose cool yet fleshy texture is said to have the consistency of human fat.[7] Hella Jongerius's "Pushed" washbasin is made from soft polyurethane. Whereas most bathroom fixtures are hard to the touch, this soft appliance is eminently suited for use in cramped spaces where flesh is often exposed.

Latex is a material that combines clinical functionality with eroticism. Matthieu Manche has created latex garments that propose links among multiple wearers and the proliferation of body parts. He uses this material of self-protection to suggest the merging and elaboration, rather than the separation and containment, of bodies. Tonita Abeyta's "Sensate" is a line of latex undergarments – some equipped with built-in male and female condoms – that aim to transform the tools of sexual hygiene into alluring fashion objects. Where love and fear are necessary bedfellows, the plush, dimly lit boudoir gives way to the bright, wipeable surfaces of the laboratory and lavatory.

Such works of brave new organicism reflect the convergence of nature and technology, information and biology. In the 1990s, plants and animals with altered DNA were dispersed through the global food market. The successful cloning of a sheep in Scotland

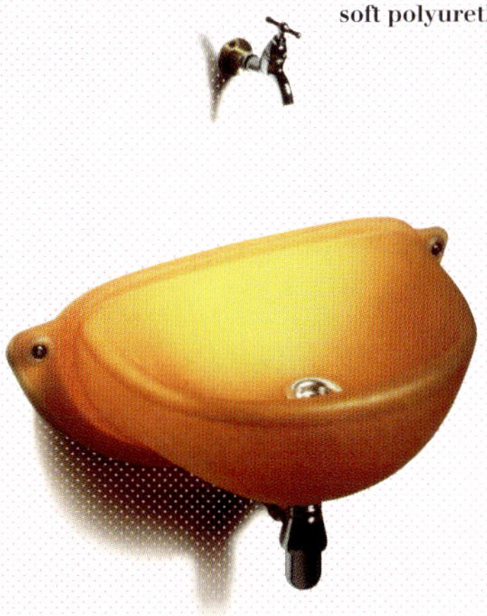

Hella Jongerius: "Pushed", washbasin, soft polyurethane, 1996

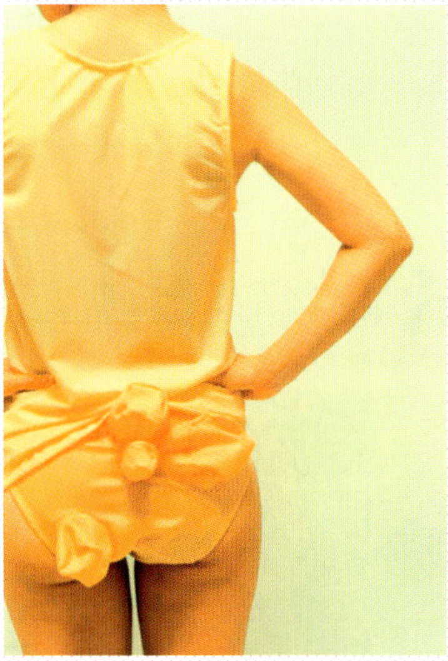

Matthieu Manche: "Fresh", latex vest, 1998–2000

in 1997 plunged a science-fiction fantasy into practice. In the summer of 2000, the human genome was mapped – a competitive venture between government and private enterprise – laying bare new terrains for medical science and economic conquest. In the mid-1990s, the new field of tissue engineering emerged, charged with the manufacture of human organs. Today, while complete hearts, lungs, and kidneys cannot yet be generated from living cells, skin is a viable medical product, grown in laboratories.

One such product is Apligraf, a cellular, bi-layer living skin substitute. It is grown in a petri dish on a bed of collagen that is seeded with living human skin cells. (The cells are harvested from the circumcised foreskin of an infant boy's penis; this neonatal tissue is ideally suited for growth.) This single human harvest can generate thousands of skin grafts. The product is shipped via FedEx and has a shelf life of five days. Apligraf is used to repair venous leg ulcers and diabetic foot ulcers, two extremely common ailments.

During the 1990s, cosmetic surgeries and products were marketed to an expanding and unembarrassed public. A host of new procedures were added to the battery of surgeries for cutting, lifting, and tightening the skin. Injections of bovine fat or collagen are used to temporarily fill shallow lines and acne scars. Chemical peels remove the outermost layer of the epidermis, erasing sun damage and other blemishes by exposing a fresh layer of cells. Injections of toxins ease wrinkles by selectively paralyzing the facial muscles.

A sense of horror as well as enthusiasm accompanies these developments. Environmentalists warn against an ecosystem unhinged by genetically altered species, while bioethicists fear a society dominated by a physically perfected, self-replicating elite. Scientific research is increasingly motivated by profit and loss – life-saving drugs and

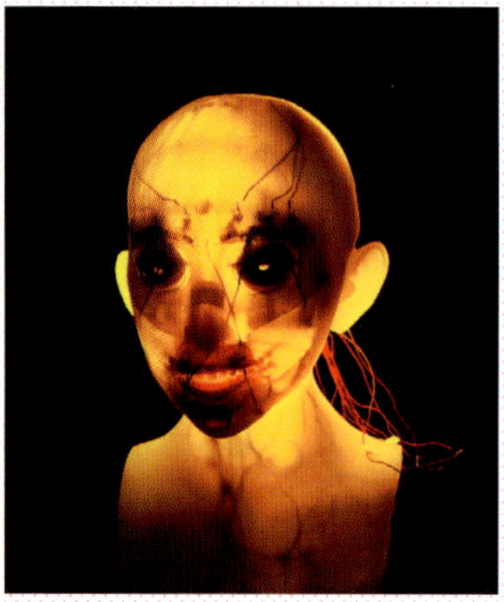

Fumio Hara and Hiroshi Kobayashi,
Science University of Tokyo (engineers):
"Second-Generation Face Robot", 2000

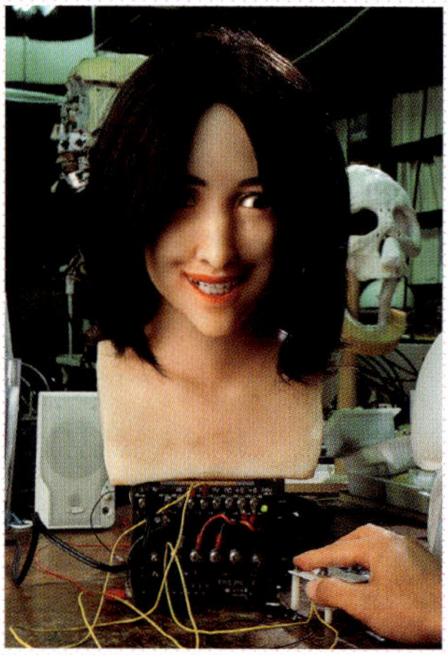

Next-generation robots are meant to respond to smiles with a smile, with a frown to a furrowed brow, and to be able to link six basic emotions in real time. Fumio Hara speaks of an "active human interface".

newly identified DNA sequences are patented and sold for financial gain, while unchecked diseases devastate local and international economies.[2] Cosmetic surgeries are consumed by the privileged few – anxious to extend the appearance of youth into lives prolonged by medical services – even as the lack of basic health care and sanitation shortens the life spans of millions around the world. Robots, ordinarily built for military and manufacturing purposes, are now being designed to express and respond to human emotions; they are planned for use in convalescent homes and other settings in which genuine human contact is costly.

Designed objects and spaces inhabit this rapidly transforming arena, where surging fears and ambitions fuel scientific discovery and stimulate the creation and consumption of new technologies. The rise of digital tools and media has changed the practice of design, providing tools for making objects and buildings that resemble living creatures – modeled with complex curves and forms – while remaining distinctly alien. This post-human organicism has taken shape most aggressively across the surface of things, as primacy of the skeleton has given way to the primacy of skin.

Architects Sulan Kolatan and William Mac Donald are developing a material called INVERSAbrane with DuPont Industries. This carbon-fiber-reinforced, organo-polymeric composite is fire-resistant, self-cleaning, and anti-microbial. Formed into a complex, three-dimensional surface, INVERSAbrane has the potential to revolutionize building, as seen in Kolatan's and Mac Donald's Resirise prototype, whose undulating skin is pocked with scoops and bladders at varying scales. These deformations in the skin serve to minimize wind loads through deceleration, diffusion, and local asymmetries. They also collect and recycle air, water, and light in order to generate micro-climates within the building.

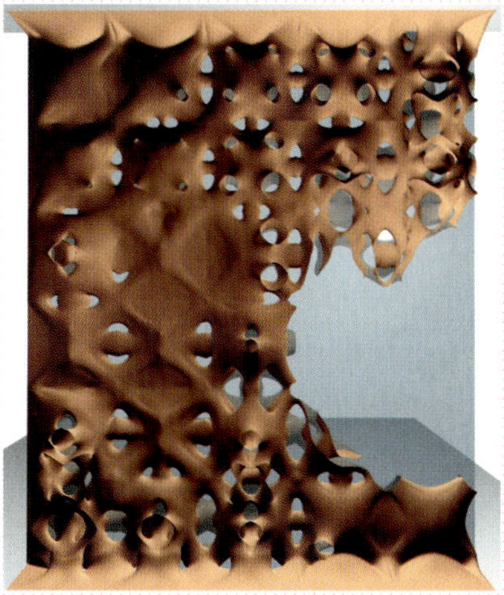

Sulan Kolatan and William Mac Donald, Kol/Mac Studio, New York: "INVERSAbrane", architectural model (waterclear stereolithograph), 2006

›
"Glory" and "Interior #6", Foto, C-Print, 2000. Photographers: Aziz+Cucher, New York City. Aziz+Cucher create digitally modified photographs of objects and spaces covered in skin.

Exploring interior walls as living, multifunctional surfaces, designer Erika Hanson has been developing new uses for Maplex, a remarkably strong, light-weight industrial material made from compressed, unbleached tree fibers, traditionally used for industrial applications. Hanson is creating mass-produceable wall panels that are cut, slit, or perforated, inviting touch or allowing light to emanate from behind them. Other innovations in wall materials include Panelite, which serves as both a finished skin and a structural material. Panelite's structural honeycomb core makes it strong and stiff yet light-weight. This translucent material, available in numerous colors and textures, has been used by leading architects, including Rem Koolhaas, for both interior and exterior walls.

Terrence Riley describes how substances of partial transparency serve to slow down and materialize the passage of light in his 1995 book *Light Construction*. Contemporary architects have exchanged the transparent skins of early modernism for physically present, semi-opaque surfaces. Buildings are clothed in multiple layers that trap and reflect light, from translucent marble to double thicknesses of glass, creating a sense of delay, a thickening of light and space, in place of instantaneous immediacy.[3]

Greg Lynn captures time by stretching and bending the skin of architecture. According to his concept of "animate form", digital design tools plunge three-dimensional structures into a space that ripples with currents of force. Lynn writes that in place of a neutral abstract space, "the context of design becomes an active abstract space that directs form within a current of forces that can be stored as information in the shape of the form." The undulating skin of Lynn's Kleiburg Block in Amsterdam records the object's passage through fields of pressure. The building exposes time in another way

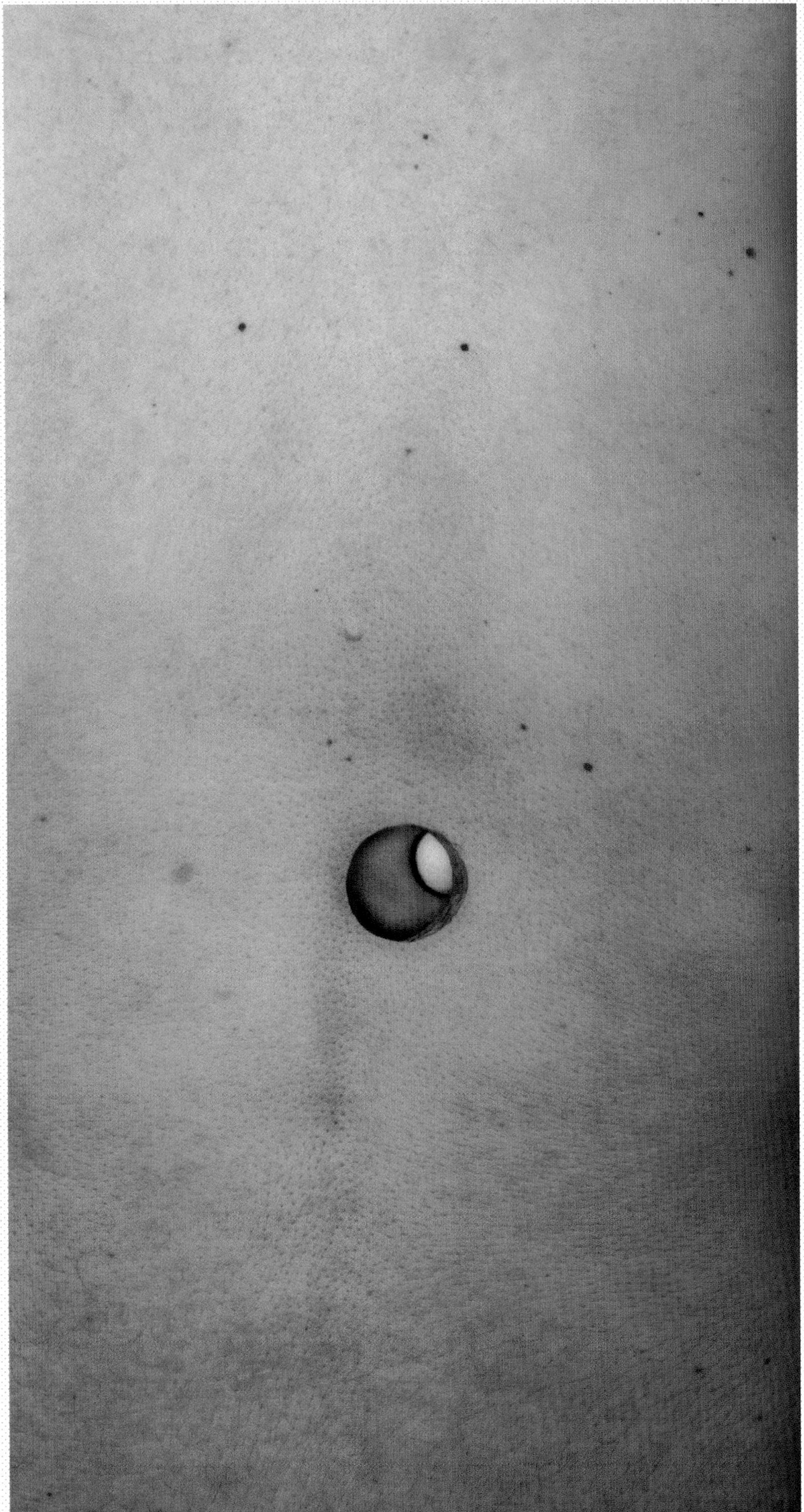

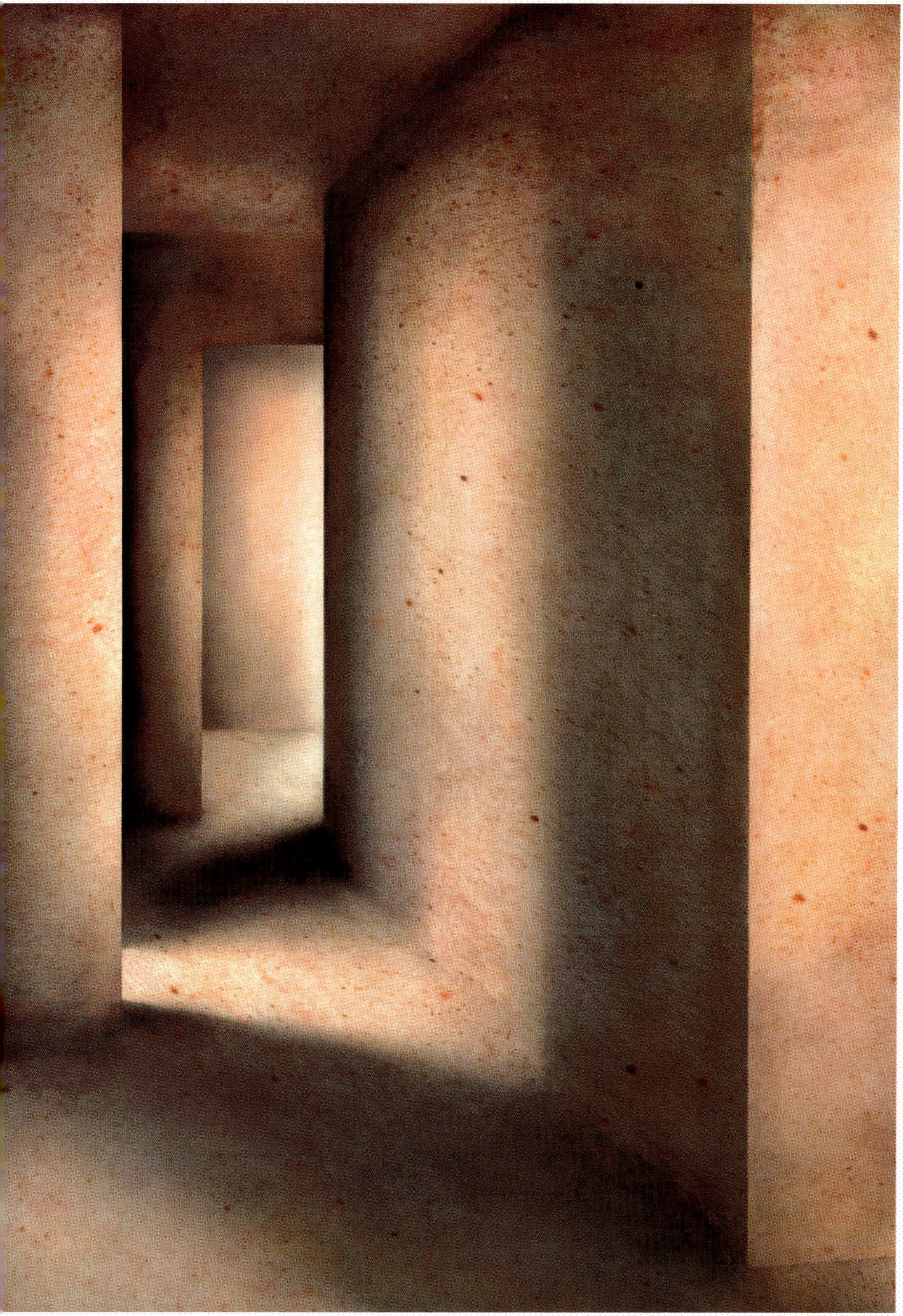

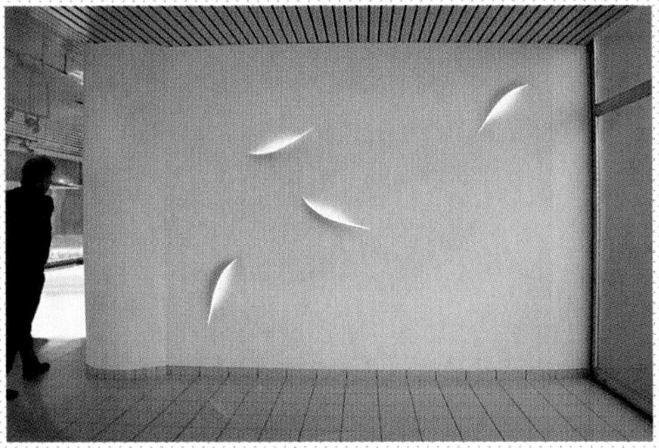

"Objects for Walls", Maplex wall system with lihgting components, 2006. Designer: Erica Hanson, manufacturer: EHV Weidman, St. Johnsburg, VT, USA.

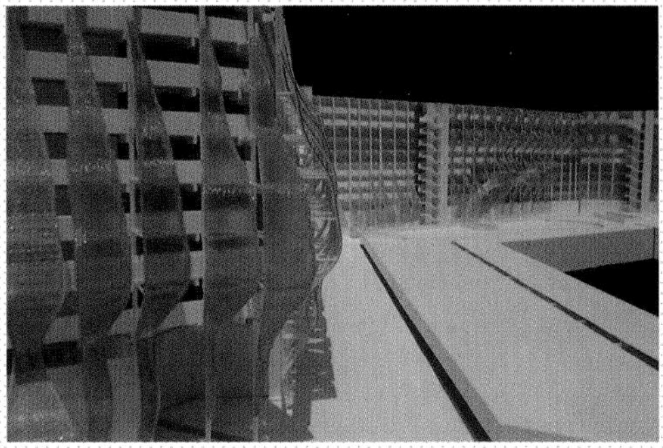

Gregg Lynn FORM: Kleiburg Block competition Architectural model (styrene, aluminum woven, acrylic), 2000

by transporting visitors through its core by way of escalators, creating a passive, cinematic experience of spectatorship. It is the first residential building to make substantial use of escalators.

The Berlin-based architecture group realities: united, led by Jan and Tim Edler, transformed the curved blue envelope of the Kunsthaus Graz (designed by Peter Cook and Colin Fournier with Spacelab.UK) into a giant media screen. The BIX media façade consists of 930 ring-shaped flourescent lights installed inside the building's translucent skin. A crude, low-resolution image forms as the lights dim on and off to varying degrees. The display system was conceived and built as an integral part of the architecture, allowing images to pulse from within the body of the building. In one sequence, huge, crudely rendered hands appear to rub and pinch the bulding's curving, bulging midriff.[4]

The use of dots of changing size or intensity is the principle governing the halftone image, which translates shades of gray into a pattern of larger and smaller spots of pigment. The deployment of this graphical device by architects and product designers reflects the interplay of two-dimensional and three-dimensional design, fields that have come to share digital tools and a communications agenda. Architecture Research Office (ARO) in New York City created a series of panels to filter light and provide changing views inside a Mahattan apartment. Peforations in each panel shift across its surface, creating a skin with fluctuating degrees of transparency. Areas of greater openness favor views from specific orientations, which relate to the design of the apartment. The panels are produced with a computer-controlled three-axis router on medium density fiberboard. The designers created numerous small-scale study models, using equipment in their own studio, to develop the final patterning.

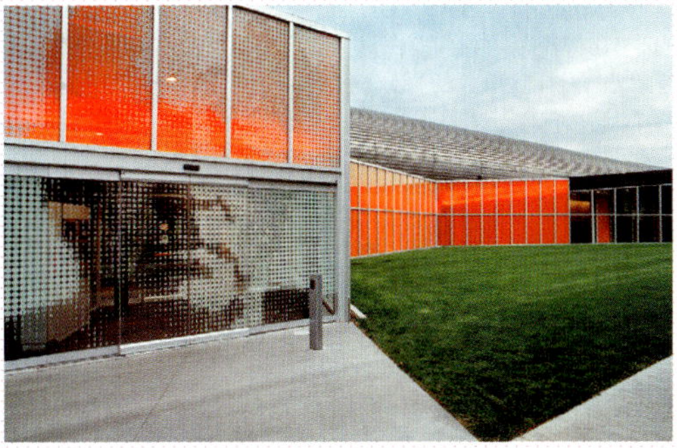

Office for Metropolitan Architecture (OMA): IIT McCormick Tribune Campus Center, Chicago, IL, completion, 2003, external glazing with IGU Panelite Insulating Glass Unit, specially produced with an inset layer of color.

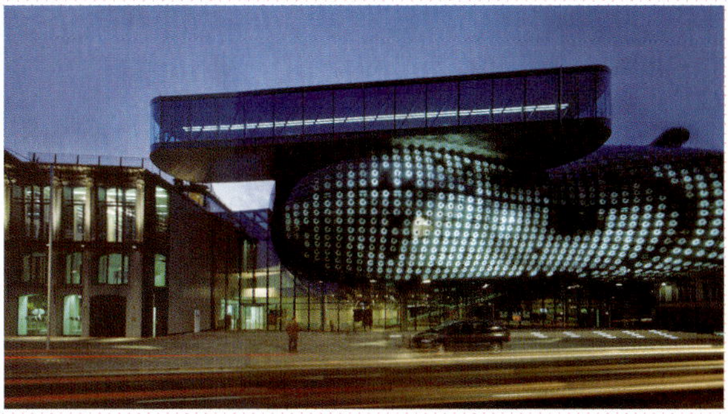

Peter Cook and Colin Fournier with Spacelab.UK, architecture Jan Edler and Tim Edler, realities: united, media façade Kunsthaus Graz, BIX media façade, 2003

Similarly, the perforations in the seat and back of Werner Aisslinger's "Nic chair" shift in size, creating relative degrees of transparency (as well as a varied tactile effect). Designer Jan Melis has created experimental packaging for Dutch Jenever, a distilled beverage traditionally sold in a brown ceramic jar. Using rapid prototyping manufacturing and software, Melis rendered a CAD image of a standard bottle in 3-D at low resolution, so that the tool paths used by the software to describe the shape of the bottle become perceptible to the eye and hand. Melis has thus exposed the graphical language that underlies the description of three-dimensional form in digitally-driven design and manufacturing.

Skins mediate between users and products in the digital realm. The customized buttons and controls found in computer interfaces are known as skins. Thousands of skins are exchanged via the Internet, allowing users to upload and download interfaces that are colored, coded, themed, and branded to suit individual whims. Avatars, the graphical icons used to represent players in computer games, are also known as skins – they are the digital surfaces of invented personae. In 2001, the first feature film cast entirely with digital, photoreal human characters – *Final Fantasy* – was released. Such characters, which consist of wireframe structures wrapped with digital surfaces, are feared by some actors as the death knell of the flesh-and-blood performer. The cyborg player is set free from time and imperfections – and requires no salary or profit share.

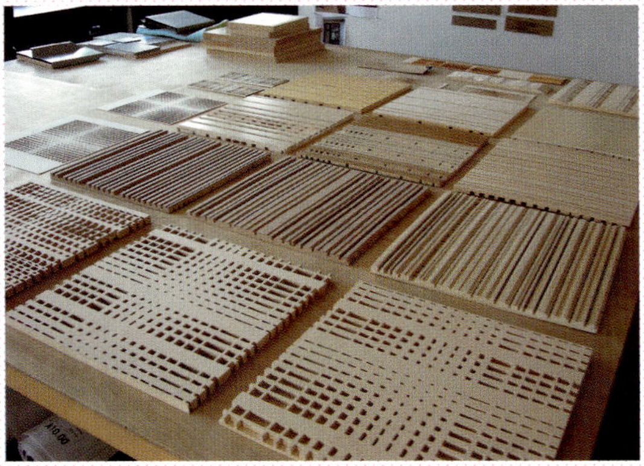

Adam Yarinsky and Stephen Cassell, Architecture Research Office LLC, New York: CNC fiberboard with a grid lattice, CNC (Computer Numeric Controlled) MDF board, 2005

The substance of the body is under renovation. The arsenal of drugs, vaccines, and mechanical replacement parts developed during the 20th century has been joined by the engineering of flesh itself. While living skin has become a commercially manufactured product, objects and buildings have come to resemble natural organisms. The barriers between body and product, self and other, nature and technology, are folding inward. The dense, luminous surfaces of contemporary objects – pulsing with hidden intelligence or taut with potential life – can be beautiful and disturbing, divine and grotestque. These industrial skins may be incubating something alien. They could be shielding us from invisible dangers or harboring the nascent growth of a predatory being. Everywhere, prophylactic skins slip into the space between people and things, forming seductive planes of contact as well as protective barriers, screens where image replaces tactility or where touch triggers a visual response – points of no entry or no return.

This essay revises and extends the essay "New Design Organics," in: *Skin: Surface, Substance+Design* (New York: Cooper-Hewitt, National Design Museum and Princeton Architectural Press, 2002).

1 Taylor, Mark C.: *Hiding* (Chicago, 1997), p. 13

2 On the commercial development of the human genome, see Wil S. Hylton: "Who Owns this Body?", in: *Esquire*, June 2001, pp. 103–11. On the ethics of cloning, see Dr. Ian Wilmut, "The Rights and Wrongs of Cloning Humans", in: *Twice* 2:1 (1998): 3-1-5

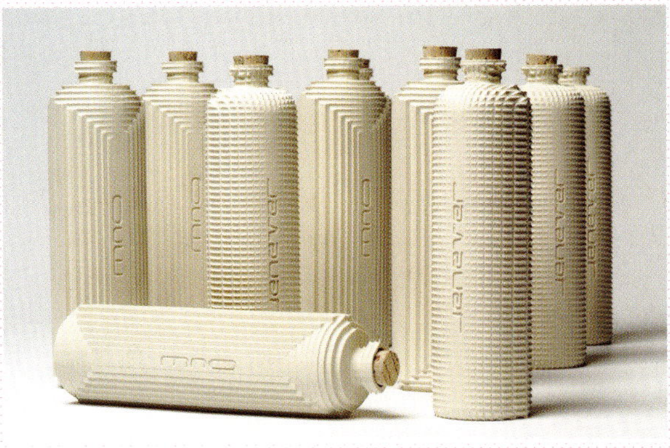

Jan Melis and Ben Oostrum: "Lo-Res" bowls, stoneware, silicon and rubber, 2003

3 Lynn, Greg: *Animate Form* (New York, 1999), p. 11
See also Joseph Rosa: *Folds, Blobs and Boxes: Architecture in the Digital Era* (Pittsburgh, 2001)

4 See Peter Hall: *Living Skins: Architecture as Interface*, 2006, http://www.adobe.com/motiondesign/main.html

TECHNOLOGICAL INNOVATIONS HAVE A DIRECT IMPACT ON INSTRUMENTS OF COMMUNICATION. FROM MAINFRAMES TO MP3 PLAYERS, IT IS NOT ONLY THEIR APPEARANCE THAT CHANGES – EVERYDAY FORMS OF COMMUNICATION ARE BEING CONTINUOUSLY TRANSFORMED AS WELL. IN HIS ESSAY ON "LIQUID MODERNITY", SOCIOLOGIST ZYGMUNT BAUMAN WRITES THAT "THE NEAR-INSTANTANEOUSNESS OF SOFTWARE TIME AUGURS THE DEVALUATION OF SPACE". NOW ARCHITECTS, LANDSCAPE PLANNERS AND DESIGNERS ARE USING NEW TECHNOLOGIES TO REVITALISE ABANDONED TERRITORIES.

--- SPACE AND COMMUNICATION ---

JOCHEN EISENBRAND – ALEXANDER VON VEGESACK

------- OPEN HOUSE -------
INTERFACES BETWEEN ARCHITECTURE AND INFORMATION TECHNOLOGY

For all the debate about how urban space is changing in the digital age, despite the *Telematic Town* and *City of Bits* and not withstanding the furore surrounding *Blob Architecture*, the buildings that surround us and in which we spend our lives are still structures made of stone, steel, concrete and glass. Like radios, telephones and televisions before them, PCs, Internet connections, cell phones and WLAN have moved into our homes without making any major changes to their outer covering. And yet don't they already exist, the interfaces between architecture and information technology? Where are they? And which direction should their development take when it comes to private homes? These questions constitute the starting point of the *Open House* exhibition and the architectural projects it presents. The participating architects were invited by the Vitra Design Museum, Weil am Rhein, and Art Center College of Design, Pasadena, California, to develop their visions for living in the next 25 years. By integrating new technologies and materials, their concepts were to offer solutions to four of the basic requirements of contemporary living space: connectivity, flexibility, wellbeing and sustainability. These are the leitmotifs of the exhibition. The examples selected for this essay will be discussed and considered in relation to some typical works of contemporary architectural practice.

Since the debate surrounding the integration of information technologies into our environment is dominated by the concepts of *ambient intelligence* and *ubiquitous computing*, these terms are discussed below by way of introduction.

The Vitra Design Museum and Art Center College of Design, Pasadena, California, invited fifteen upcoming architectural firms and designers to develop their visions of home living for the *Open House* exhibition. The projects show some alternatives to the concept of the intelligent house, today's dominant vision of tomorrow's living. As a rule, intelligent houses confine themselves to lighting scenarios, sound systems or automated temperature control. Environments that adapt independently to the user and his needs represent a greater challenge. The model for such environments is "ambient intelligence", which works on the assumption that the computer will become part of our environment and functional objects will be networked, communicating with one another via radio (RFID tags). Accordingly, the concept predicts that the computer as a single device will disappear, to be replaced by things

AMBIENT INTELLIGENCE: SEAMLESSNESS AND PERIPHERAL AWARENESS

For a good ten years now, research and industry have been propagating the intelligent environment, made up of networked components and capable of interacting with humans, as the future of home living. One of the key features of this *ambient intelligence* is that it is embedded in the environment, that its networked components are integrated into both the space itself and the things we surround ourselves with. *Seamlessness* has become a popular catchword for describing this idea. So far "embedded" has also meant adapted to the respective environment or, to put it another way, with no visible consequences for the architecture – as demonstrated by the conventional coverings of the intelligent houses currently promoting networked living around the globe. "Intelligent" means that the system can react to certain situations. In order to do so its components, its sensors, must recognise not only the user but the respective situational context as well. This in itself is difficult enough and requires the system to be customised to its user. But there is an even greater challenge: *ambient intelligence* is supposed to be adaptive, it must be capable of changing the way it reacts to the user, his characteristics and daily activities and adjusting itself to his needs. Thus – or so the pipe dream goes – such systems should ultimately be able to anticipate the user's needs, without him having to communicate them explicitly.[1] Advocates of this concept maintain that people have adapted to the computer long enough to learn how to operate it; now, they say, it is time for the computer to adapt to us and learn to communicate in a way that corresponds to human nature.

In the intelligent show houses, however, the talents of the technical equipment have so far been confined to pre-programmed lighting scenarios, sound systems or automated temperature control. Some of these houses come complete with a *personal digital assistant* that functions as a central remote control for all the appliances and technical

that communicate both with each other and with us. But are such things any easier to handle than current household technology? With examples taken from both the *Open House* exhibition and reference projects in the field of contemporary architecture and design practice, the authors show three areas in which architecture and digital worlds are entering into alliances. Kinetic, responsive walls react to direct stimuli like touch, noise and light, textile building coverings can be reconfigured using pneumatic muscles. Other concepts link architecture with the Internet: an epoxy-resin tower that interacts with the local inhabitants and uses a mix of coloured light to reflect their changing moods. Timesharing is to make optimal use of residential towers with small private units and large, networked communal areas. Finally, another approach being pursued by designers and architects alike is to draw our attention

equipment. Yet a truly intelligent environment would no longer require such a single device. Instead the interaction would function by means of motion sensors, pressure sensors in the floor, heat sensors or microphones and voice recognition software. The occupant could be located by means of a transmitter kept on his person. Real-time gesture recognition and processing is another area of research that could contribute to the creation of intelligent environments. This would however require a system that can recognise objects in the house that are not themselves active components of it – they would have to be fitted with Radio Frequency Identification tags. All in all, an extremely complex technology that is still far too expensive for widespread use.

Over the last ten years *interaction design* has established itself as a discipline in its own right and is taught and developed at institutions such as the Royal College of Art in London, the Interaction Design Institute in Ivrea, Italy or the Interactive Institute in Stockholm. *Interaction design* operates at the interface between design and computing. Its goals include the development of solutions for intelligent environments.

A host of research projects have sprung up in the wake of *ambient intelligence* in recent years. It is a phenomenon not only propagated by industry, but under investigation by a number of reputable research institutions as well.

One of the basic assumptions for belief in the realisation of *ambient intelligence* is that communication or interaction between a human being and his networked environment can function not only at a direct level but by means of peripheral awareness as well.[2] Although this primarily refers to awareness on the fringes of our field of vision, the term is also applied in a general sense to all the impressions and information that we only register and process at a subconscious level, without concentrating on them. This could be something concrete like weather conditions or a less tangible phenomenon such as a certain mood. In the last few years interaction between man and his intelligent environ-

to invisible radiation, data and energy streams, or make them perceptible to the senses. This type of project might involve extreme manipulations of light, indoor temperatures and humidity or building coverings which, like the human skin, store water and compensate for temperature fluctuations. In contrast to the intelligent houses of today, *Open House* also pleads the case for sustainable architecture that encourages social interaction and opens living space up to the outside world.

ment based on peripheral awareness has been the topic of several research projects dedicated to finding out how we can register information out of the corner of our eye, as it were, via changes in the things that surround us, instead of turning to the computer, television, radio or telephone as we usually do. One of these research projects at the MIT in Boston resulted in the founding of Ambilight. The company develops lamps that are connected to the Internet and can be programmed to change colour in response to a chosen set of facts such as the weather or stock exchange index. They are intended to provide discreet, indirect information – by means of peripheral awareness, they let the user know how cold it is outside or how his shares are doing. *Calm technology* is a phrase often used in this context to refer to technology that supposedly steps into the background rather than imposing itself upon us. What it actually does, however, is add another information level to our environment.

In this respect the concept of peripheral awareness is slightly reminiscent of Muzak, the musical background sound invented by the company of the same name (founded in 1934). From the 1950s on Muzak, also known as elevator music, became a widespread phenomenon. Behind the glass facades of air-conditioned buildings, it created an artificial acoustic atmosphere which, it was claimed, could either soothe the occupants or inspire them to throw themselves into their work – without drawing attention to itself and becoming a distraction.

We often only perceive the activities of the people we live with at an incidental, peripheral level as well. Without giving it conscious thought, we know where the other person is and have some idea of what he is doing. This was the basis for a whole series of research projects dedicated to investigating how, when people are separated, information technology can be used to establish closeness at a distance and simulate togetherness. At the Media Lab Europe in Dublin, which was forced to close in 2005 due to financial difficulties, these projects were grouped under the motto *intimate remote awareness*. Some took the form of networked furnishings: two kitchen tables in two different apartments were fitted with an RFID tag reader and a video projector and connected to the Internet. The table in one of the apartments registered when an RFID-tagged object was placed on it and transmitted the information to the other table in real time, via a permanent Internet connection. An image of the same object was subsequently projected onto the surface of the second table. A system which, like other, similar projects, requires a 24-hour online connection to keep the separate-yet-connected protagonists up-to-date on one another's daily activities.

The *Digital Family Portrait* developed at the Georgia Institute of Technology as part of the *Aware Home Initiative* is based on a similar principle but uses a one-way connection. On the transmitter side, the activities of a far-off family member are registered and conveyed to a picture frame on the receiver side. The frame is connected to the Internet and contains a portrait of the person whose activities are to be registered – and who therefore needs to be equipped with a transmitter. The information conveyed is restricted to a superficial level, such as which rooms the person has been in and how much he has been moving around the apartment. The idea is primarily aimed at the elderly and would ensure the recipient of the information has a permanent if only approximate idea of their relative's health.

The interactive cushions conceived at the Interactive Institute in Ivrea take a similar approach. They come in pairs and are connected over a distance by means of a transmitter and receiver. Touching and pressing one cushion changes the patterns and colours of its opposite number, again with the aim of simulating physical closeness.

Essentially the projects described here, which are typical examples of an independent field of research, are dedicated to the question as to how the computer can become part of our everyday environment. A text by Marc Weiser, published in *Scientific American* in 1991, pointed the way ahead for such ideas. It was in this article that the IT researcher from the Xerox Parc Research Institute in Palo Alto, California, coined the term *ubiquitous computing*. According to Weiser, the computer should no longer stand alone in the home or office as a solitary unit but should be replaced by enhancing all the things we deal with on a daily basis with digital technology. This would, Weiser assumed, result in the computer as a single device becoming superfluous and eventually disappearing. His thesis: "The most profound technologies are those that disappear. They weave themselves into the fabric of everyday life until they are undistinguishable from it."[5] Here Weiser was referring explicitly to Sigfried Giedion's expositions in *Mechanisation takes Command*. 50 years before Weiser, Giedion had looked back on the invasion of mechanical and electrical devices into all areas of life and investigated the way they had changed our everyday existence. According to Giedion, it was the small electric motor found in every electrical kitchen appliance that played a major role in changing the household whilst nevertheless remaining invisible.

The major difference between the invasion of the household by electrical appliances and the vision of *ambient intelligence* is however the distinction between a single device and a network, between appliances that are only active for a limited period of time and a permanently active environment – or between appliances that permit the user to decide when they are needed and a system that intervenes in our lives and relationships with virtual autonomy. Furthermore, the central question appears to be whether handling today's highly miniaturised and thus mobile computerised devices has not already become second nature and whether we will not always find this type of device easier to deal with than an intelligent environment.

Without referring to such things directly, Weiser's vision seems to have been influenced by *virtual reality* scenarios in which a person moves between apparently three-dimensional scenes and can interact with the space surrounding him in various ways. In virtual space, such as the scenery generated by computer games, surfaces can be active and dynamic and thus constitute the interface to a new level of the game, to a different dimension.

DYNAMICISING THE WALLS

In the projects described above it was still a case of individual, networked products and furnishings which only achieved the ideal of being embedded in the environment with respect to their appearance: they did not really become a physical part of the wall. However, the most visible interface between information technology and architecture and the most obvious embodiment of *ambient intelligence* are the kinetic, responsive walls designed and realised by various architects and research facilities in recent years. One of the first and probably best-known projects is the "Aegis Hyposurface" by Decoi, for whom Marc Goulthorpe developed a faceted metal surface measuring 8×8 metres. It consists of 1,000 small triangular plates moved by a total of 896 pneumatic pistons.

These agile plates are connected to a substructure via a kind of telescopic finger that allows their individual position to be altered by up to 50 cm within a fraction of a second, at a maximum speed of 60 km/h. They are controlled by means of a matrix of electronic sensors linked with a computer.

Another, comparable project, the "Super Cilia Skin" by the MIT Tangible Media Group, involves a computer-controlled membrane. A thick field projects out of its surface, made up of threads of equal length that respond to touch but can also be controlled by a computer.[4] The name Cilia refers to microorganisms like the paramecium, which moves by means of the eyelash-like structures (cilia) covering its body. The movement of the "lashes" is registered and can be transmitted to a second wall somewhere else, the lashes of which then execute the same movements. As a kinetic wall, the "Super Cilia Skin" thus picks up on the same concept of simulating closeness at a distance described above.

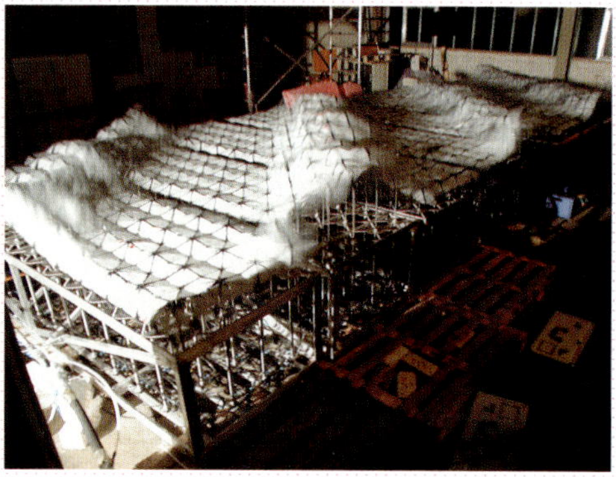

Marc Goulthorpe/Decoi, "Aegis Hyposurface", 1999–2001
Faceted metallic surface that changes shape three-dimensionally in response to electronic stimuli. A total of 896 electronically controlled, pneumatic pistons create dynamic landscapes in real-time.

In the *Open House* exhibition, the "Parcel" installation by Scandinavian group Krets presents a kinetic wall that is animated and moved by background noises. It is made up of modules that can not only be combined as required but are compatible with mass-production as well. The name, Parcel, refers to the manufacturing method: the modules are punched and folded by means of the same technology used in the packaging industry. Microphones, circuits and LEDs are integrated into the modules, enabling the structure to respond to background noises with light as well as movement. Following the initial prototype presented in Stockholm, the *Open House* prototype Spline Graft will be made of so-called shape-memory metals, which can be deformed but return to their original, "remembered" shape after heating.

Dutch architect Kas Oosterhuis of ONL goes one step further, from the deformable, two-dimensional wall to deformable, three-dimensional space. His *Muscle* project involves a flexible textile building covering that can move in any direction. It is deformed by the relaxation or contraction of 94 pneumatic muscles, which can either be manipulated by 'remote control', i.e. via a computer, or triggered in response to direct interaction, by means of motion sensors integrated into the covering itself.

In the follow-on project at the Technical University of Delft, *Muscle Reconfigured*, Oosterhuis conducted an experiment that emphasises the deformation of the internal spatial boundaries. Under his direction, the university's *Hyperbody Research Group* created a kind of shell made of strips of ductile sheet metal and shaped like a zero lying on its side. The metal plates, reinforced on the floor with narrow transverse strips of foam, were again connected by means of pneumatic muscles, enabling them to be bent into seating or lounging elements on the floor.

The rapid spread of such projects in recent years is no doubt largely due to very pragmatic aspects. Only now are the technical means, the appropriate sensors and actu-

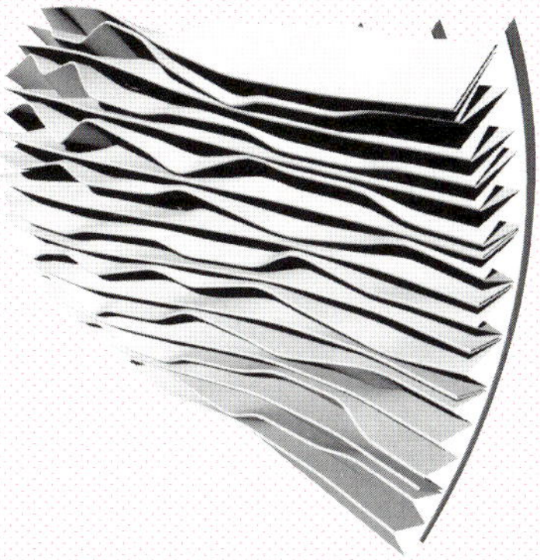

Krets Research Group, study on the shape and behavior of the new "Spline Graft" prototype developed for the Open House exhibition, 2006

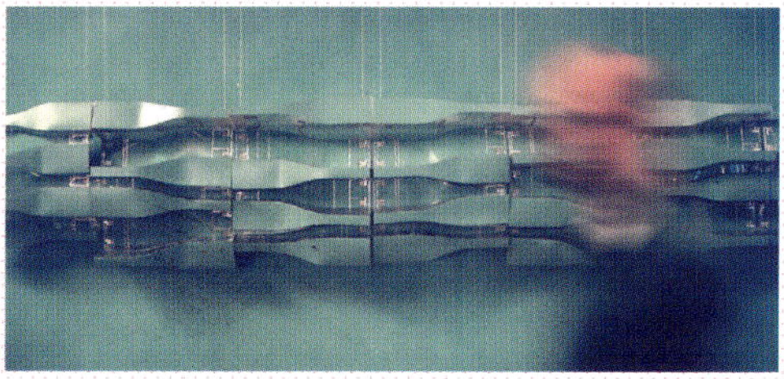

Krets Research Group, "PARCEL", a responsive wall paneling system with integrated sensors, lighting elements and loudspeakers, first went on show in Stockholm's Concert Hall during the Stockholm Arts+Science Festival, 2004.

ators like the pneumatic muscles available. The distant ideal that such experiments strive for is space so flexible that it can react to our individual spatial needs, expanding ad hoc wherever extra room is required and shrinking where it is not needed.

And yet projects like this also demonstrate just how much technical effort is still required before we can give real shape to the type of space that can be depicted so simply and convincingly on a monitor. Nor does there seem to be any immediate solution to the challenges of acoustic and thermal insulation, which are rarely solved satisfactorily even in the case of conventional flexible walls in residential buildings. And, finally, one more question remains: where exactly are we supposed to put the furniture? The nature of these projects thus oscillates somewhere between experimental arrangement, artistic installation and architectural prototype.

ARCHITECTURE AND NETWORKS

Linking the technology and logic of the Internet with buildings represents another feasible interface between architecture and the digital world. One of the most interesting projects to be realised in this context is the *D-Tower* (1998–2004)[5] by Nox Architects and artist Q. S. Serafijn. The project, which originated in Doetinchem, Netherlands as part of a programme to promote art in public spaces, aims to establish a network between the inhabitants of a community and, at the same time, serves as a visible symbol of it. There are three elements to the project: the 12-metre-high, organically shaped tower, which is visible to all and has already been the subject of numerous publications, was designed by Lars Spuybroek (Nox). The structure is slightly reminiscent of a tooth standing on its roots and was put together out of 19 individual parts made of glass fibre-reinforced epoxy resin. This is the visible part of the project. It also includes a questionnaire distributed to the inhabitants of the district where the tower is located and a Website by the artist, which

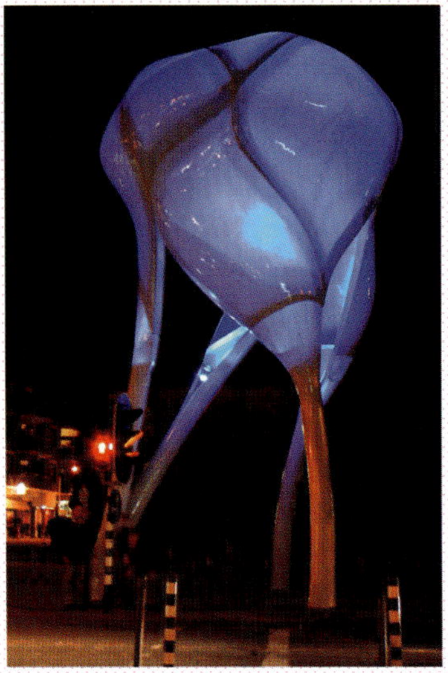
Nox, "D-Tower", Doetinchem, 1998–2004. The tower depicts feelings such as happiness, love, fear and hatred – as polled via internet from inhabitants of the Dutch town of Doetinchem. Blue stands for happiness.

residents log onto in order to answer the questionnaire. The questions are geared to determining how the members of the community are feeling at any given time. The Website analyses their answers automatically and determines the general mood in the neighbourhood. Each mood has been allocated a colour: blue for happiness, green for hatred, red for love, yellow for fear. The tower changes colour depending on which mood predominates. This is made possible by three lamps on the inside of each leg. Each lamp is made up of 30 LEDs in the RGB scale. The glowing tower, which keeps people informed about the prevailing mood in the immediate vicinity, is thus also a visible symbol of the connections, or network, between the people living in the community.

Architectural historian Beatriz Colomina gives a convincing description of how, in principle, the 20th-century buildings designed and in some cases realised as houses of the future cut themselves off from their surroundings and direct their occupants' attention inwards. From Buckminster Fuller's *Dymaxion House* and the Smithsons' *House of the Future* all the way to the habitation capsules of the 1960s, the visions of future living embodied by these *Houses of the Future* were always dominated by isolation from the outside world and concentration on the nuclear family.[6] This same observation also applies to the intelligent houses promoting the future of living today and is equally true of the *ambient intelligence* projects that establish closeness at a distance so that the user is no longer required to leave the house.

In contrast, some of the designs conceived for the *Open House* exhibition show how connectivity can be used to create community and make shared, more meaningful use of spaces. One such example is *Seoul Commune 2026: Rethinking Towers in the Park* by Korean firm Mass Studies. In South Korea, 72 percent of all households were provided with a broadband Internet connection in a period of just five years. The country is currently being equipped with an extremely fast wireless Internet service by the name of WiBro,

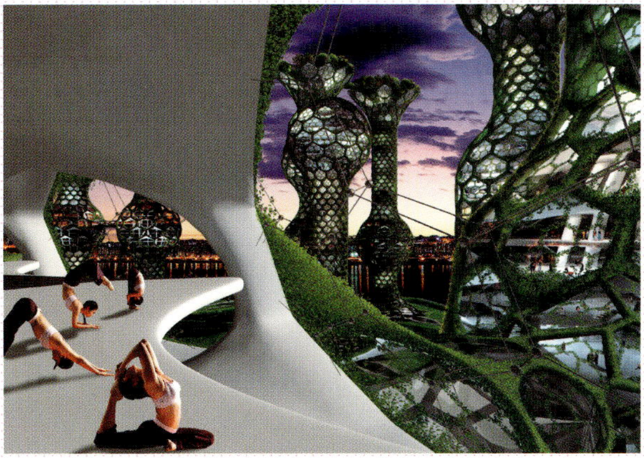

Mass Studies, Seoul Commune 2026: "Rethinking Towers in the Park", 2006, detail

›

Mass Studies, Seoul Commune 2026: "Rethinking Towers in the Park", 2006

which even permits a permanent online connection when the user is on the move.[7] In addition, Seoul is one of the most densely populated places in the world, with more than 40,000 people living in an area of just one square kilometre. A new concept has become fashionable in recent years: residential towers built in a park and offering better-quality living space than what is otherwise available. *Seoul Commune 2026* picks up on the concept of towers in the park but utilises new communications technology to create a third type of space in addition to private living space and the public space of the park. This third kind of space is used communally, which is why the private cells in the towers have been reduced to the minimum required for a bedroom and bathroom. All other rooms are in permanent, communal use or can be viewed online and reserved if needed.

Japanese architect Hitoshi Abe proposes something similar, although he prefers to make better use of the existing urban structure rather than build something new. The concept of *Megahouse*, the project he contributed to the exhibition, is based on a new timesharing system that aims to make unused city space accessible for flexible, individual usage. *Megahouse* is a totally new kind of Internet-based administration system that attempts to integrate the unused space spread throughout the city so as to maximise usage of municipal buildings and indirectly enhance quality of life in a shrinking and increasingly mobile society. The spaces in question are equipped with a special door – a so-called ZapDoor – that uses biometric scans to restrict entry to those with a booking. Besides the space users and the space providers, there is a third category of network participants that supplies the so-called space "fillings", which might either be of a material nature or take the form of services.

Both projects, *Megahouse* and *Seoul Commune 2026*, propagate a new kind of living which would however necessitate a different understanding of property and privacy. And

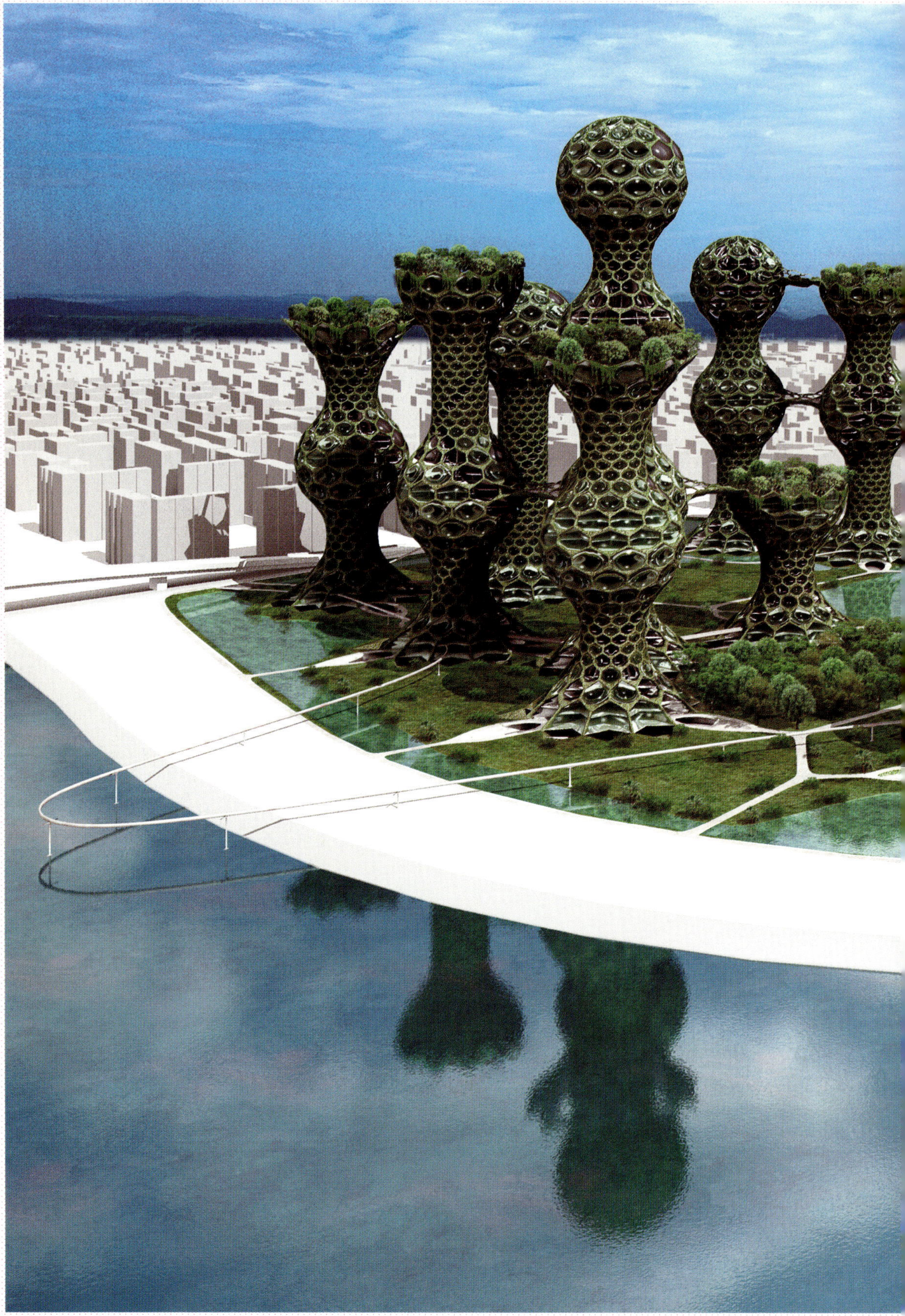

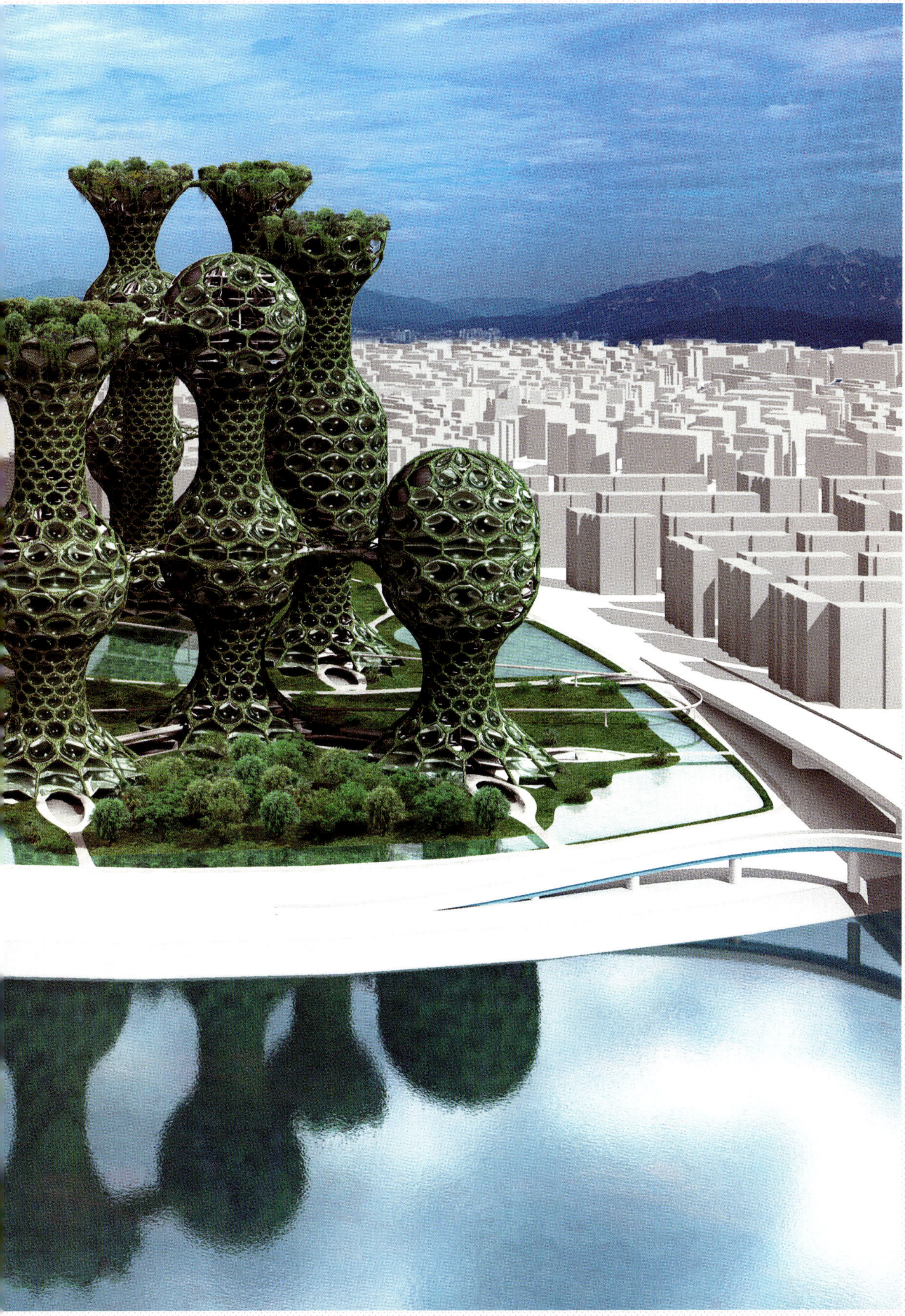

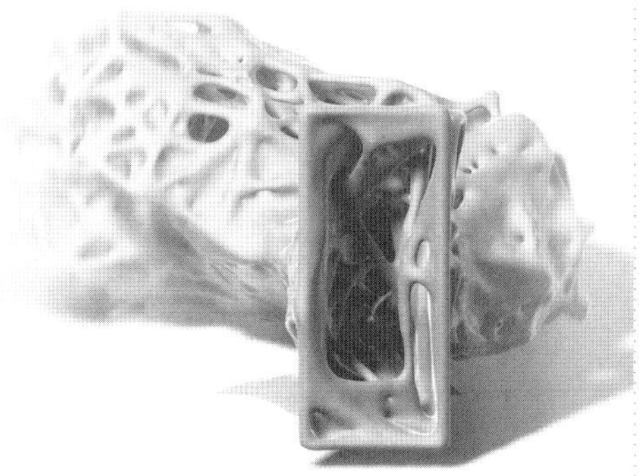

R&Sie(n), Dismiss hypnosis chamber, 2006

yet neither Hitoshi Abe nor Mass Studies seem to be under any illusions as to the fact that their projects would involve different people using spaces together. Both projects work on the assumption that the communally used rooms would be kept under surveillance by means of sensors or cameras.

The vision of urban living space that originates and grows by means of networking its inhabitants and responding to their interaction with the urban structure conceived by French firm R&Sie(n) for the *I've Heard About* exhibition at the Musée Moderne de la ville de Paris is equally radical and even further removed from reality. The urban organism devised by R&Sie(n), the form of which at times reminds you of a termite nest and at others of foam structures, also goes by the name of *I've Heard About*. The organism uses a computer-controlled machine to generate and expand itself. The apparatus responsible for its production looks like an elevator that glides along the outside of the structure, discharging a material that immediately hardens into building mass out of several tubes. This machine is indirectly controlled by impulses from the inhabitants of the structure, transmitted by nanosensors that they have previously breathed in. The structure responds immediately to the mental state of its occupants and their interaction with one another. *I've Heard About* thus translates the *Open Source* principle, which originated in computing and spread via the Internet, into architecture. Although at first glance this might seem democratic, the overall aesthetic form stipulated by the architects nevertheless represents a totalitarian entity. In order to prepare for life in this kind of structure, the firm suggests undergoing hypnosis to achieve a new level of consciousness. Accordingly, for *Open House*, R&Sie(n) conceived *Dismiss*, a hypnosis chamber that protrudes out of the wall of a conventional apartment like a tumor. The occupants can settle down inside and allow themselves to be hypnotised by film projections. Might it also be an ironic reference to the didactic moment inherent in many a design for residential architecture since the advent of modernism?

DEPICTING THE DIGITAL

Whereas the ideal of *ambient intelligence* envisages making technology invisible and many digital-age technologies are invisible anyway, designers work with the visible, they see making things visible as one of their primary tasks. The work of British designers Anthony Dunne and Fiona Raby, for instance, focuses on the social and cultural consequences of the advance of new communications technologies.[8] Both teach at the Royal College of Art, where Anthony Dunne is head of the Department of Interaction Design. The furnishings they create take an ironic slant on the subject of connectivity whilst simultaneously drawing attention to the invisible and imperceptible electromagnetic radiation and digital data streams that surround us. An apartment furnished with the prototype objects they have developed would constitute the exact opposite of the ideal of an intelligent environment, whilst at the same time questioning its implications.

The surface of Dunne&Raby's "Compass Table", for instance, contains 5 × 5 compasses arranged in a grid pattern. The compass needles swing when electronic objects are placed on the table. The "Electricity Drain Stool" cleanses anybody that sits on it of electromagnetic radiation, whilst the "GPS Table" contains an integrated global positioning sensor, enabling it to know exactly where it is at any given time. When the satellite connection is interrupted, the table posts a message on the display in its surface: "Lost!" – as if it were only any use when networked.

Dunne & Raby's "HideAway Furniture" was produced under the motto *Design for Fragile Personalities in Anxious Times*. The various pieces have a surface adapted to the texture of the parquet floor and are shaped to correspond to a certain sitting or lying position. The furniture blends into its surroundings and allows the user to escape from his home environment by embedding himself into it seamlessly, leaving the communicating appliances to their own devices.

Another British designer duo, Ben Hooker and Shona Kitchen, starts from similar premises but draws very different conclusions about living. They propose the deliberate and active medial linking of the interior with the outdoors. For the *Open House* exhibition they conceived "Electroplex Heights", a permanently changing residential structure set in the infrastructure belt around modern London. In principle, the residential complex designed by Hooker & Kitchen resembles an oil rig on stilts and is placed above the industrial zones and trading estates that surround today's urban conurbations. It is in such places, with their logistics halls and call centres for Internet retail, criss-crossed by high-voltage lines and telephone masts, in these transit zones dissected by motorways and the approach paths of nearby airports, that Hooker & Kitchen see the spatial equivalent of Cyberspace and, at the same time, a symbol of true contemporary urban living. Each of the apartment blocks is located at a fracture in the landscape, such as the point where building land merges into wasteland. Hooker & Kitchen acknowledge the perception of an industry-dominated, urban environment as an aesthetic experience, using various electronic media to process and enrich it. Images and data from the surroundings are

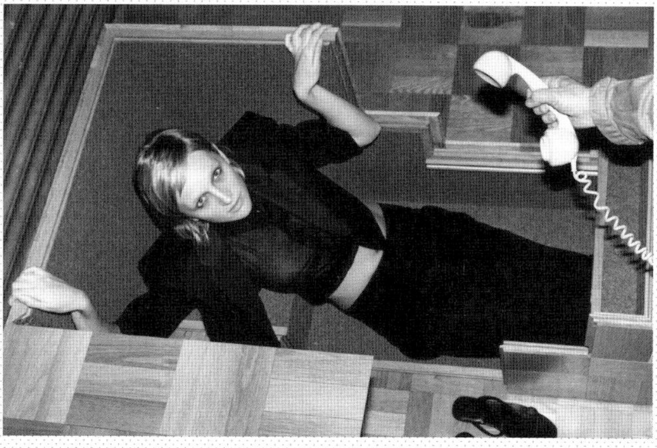

Dunne & Raby, "Designs for Fragile Personalities in Anxious Times", Hide Away Furniture, 2004

recorded by different kinds of technical apparatus and transmitted to the apartments. From time to time the platform is dismantled and rebuilt at a new topographical fracture. The crucial point: a community-building exchange is also intended to take place between the occupants of the residential units on the platforms – via electronic media. A contemporary variation on *train spotting* and *bird watching* and yet, in the digital age, also the counterpart to the galleries and lofts in converted factory buildings that came into fashion in the 1970s and 1980s and have been contributing to the *gentrification* of their once industry-dominated districts ever since. Bored by the increasingly interchangeable face of city centers that are degenerating into theme parks as a result of colonisation by global chain stores and restaurants, these "pioneering residents" regard the infrastructure belt as the actual city.

THE ARCHITECTURE OF ATMOSPHERES

"Today's urban spaces consist of three layers [...] Ignoring the first layer [which is based on the course of nature], the period of mechanical technology created a second layer of extremely geometric, homogeneous city spaces. Today there are spaces that form part of a third layer, but because of the gradual spread of electronic technology it is not visible to the naked eye. If we make clever use of this third layer, couldn't we unearth the first layer again, the layer that was obliterated by the second? This would mean using a skilful combination of the first and third layers to rediscover nature."[9]
Toyo Ito

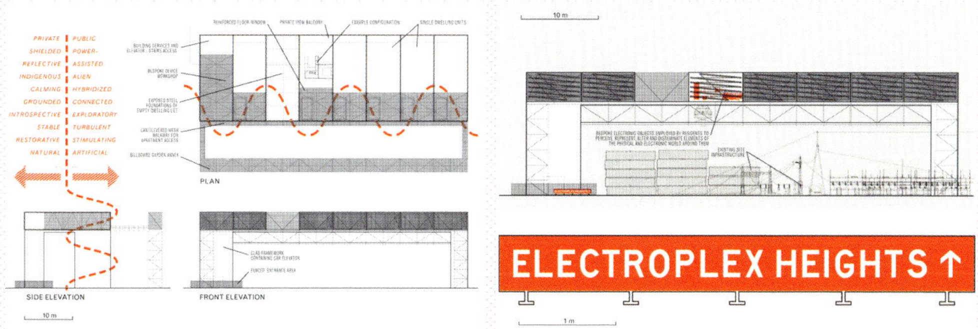

Hooker & Kitchen, "Electroplex Heights", 2006

Drawing attention to phenomena that surround us and influence our lives without themselves being visible also characterises the work of architects Decosterd & Rahm. With their "physiological architecture", they devote themselves to the "physically electromagnetic, biological and chemical interaction between architecture, the environment and our own organism."[10] Their tools of choice include light and manipulating the composition and temperature of inside air. For the *Hormonorium*, Switzerland's 2002 contribution to the 8th Architecture Biennale in Venice, they created an installation that simulates the atmosphere in the High Alps, with light as bright as in the snow-covered mountains and air with a reduced oxygen content. The *Maison d'hiver* they conceived for French artist Fabrice Hybert (2001–2005) in Vendée, France, is a house with a Tahitian climate.[11] Philippe Rahm, who has been working on his own since 2004, is currently building the *Archimedes House* and the *Mollier House*, the architecture of which is largely determined by the air conditions in the individual rooms. In the *Mollier House*, located next to a lake, the humidity created in the rooms increases as you get closer to the lake, whilst the rooms of the *Archimedes House* are staggered according to the different temperatures that prevail within them.

This architectural practice draws attention to the fact that, thanks to air conditioning, artificial lighting and increasing isolation from the outside world, modern architecture has created homogeneous climatic spaces all over the world, with a constant temperature of 21°C, relative humidity of 50 per cent and a luminous intensity of 2000 lux. By staging unaccustomed sensory experiences, the work of Decosterd & Rahm illustrates that architecture is more than concrete and brick, glass and wood, more than just floors, walls and ceilings, that it is in fact a substantial component of the spatial atmosphere it creates. This coincides with the ideas of architectural theoretician Reyner Banham, who demanded that architects themselves take control of the *environmental management* in their build-

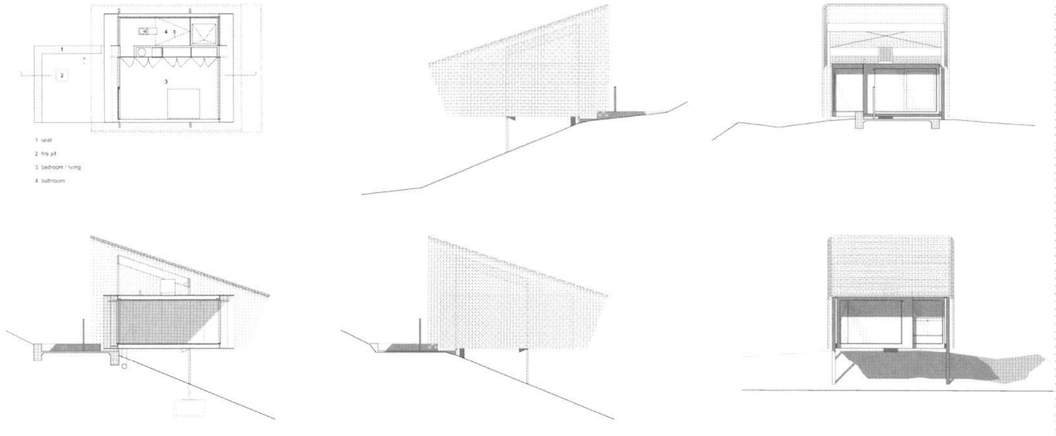

Sean Godsell, Primitive Hut for the 21st Century, 2006

ings rather than handing it over to conditioning engineers.[12] In allusion to Banham's theories, Philipp Oswalt coined the phrase "the architecture of atmospheres", meaning the energetically formed space as the opposite pole to the dominant idea of homogeneous and constructive space. The architecture of atmospheres is anything but homogeneous, it is instead a field of varying density and intensity, not delimited and isolated but part of its environment, with which it conducts an exchange.[13] Whilst not new, the associated interpretation of the skin as a filter for light, air, noise and energy flows rather than a dividing barrier between the inside and the outside nevertheless offers huge potential for future developments towards sustainable architecture that is pleasant to occupy and live in as well. Several projects in the *Open House* exhibition consider the possibilities of dynamicising the covering of a building in this way. The covering of the *Jellyfish House* by Iwamoto Scott, for instance, almost seems to be an organism in its own right. It serves as a reservoir and water filter, as a heating or cooling unit and, thanks to its integrated *phase change material*, can independently compensate for changes in temperature. This is not the only project in which natural systems serve as the conceptual model. With his *Primitive Hut for the 21st Century*, Australian Sean Godsell refers explicitly to human skin as the inspiration for his intelligent building cover: it filters light and dust, secretes coolant and uses photovoltaic cells to provide energy. For the time being, technology and architecture are not even close to copying the intelligence and adaptability of natural organisms 1:1. There is still a long way to go before they can accomplish anything that remotely resembles human skin and its numerous capabilities – including self-repair. Nevertheless, the progress being made in the field of new materials, as documented by numerous publications and exhibitions in recent years, gives grounds for hope that new intelligent possibilities and solutions will soon become reality. In contrast to what most projects on *ambient intelligence* propagate, *Open House* does not advocate enriching in-

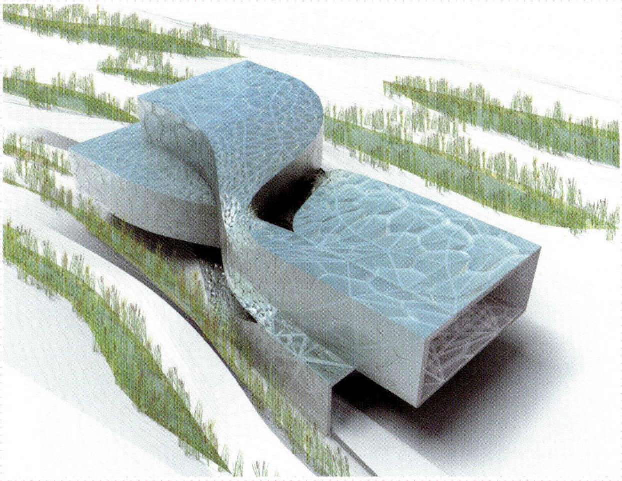

Iwamoto Scott with Proces2, "Jellyfish House", 2006

terior living space with information layers so as to effectively cut it off from the outside world. Instead, *Open House* pleads the case for opening living space up to the outside, for sustainable architecture and more social interaction.

1 Aarts, Emile: "Technological Issues in Ambient Intelligence," in: Emile Aarts, Stefano Marzano: *The New Everyday. Visions on Ambient Intelligence* (Rotterdam, 2003), p. 14

2 Wineskin, Craig et al.: "Ambient Displays: Turning Architectural Space into an Interface between People and Digital Information", in: *Proceedings of the First International Workshop on Cooperative Building*, February 25–26 (Vienna, New York, 1998)

3 Weiser, Marc: "The Computer for the 21st Century," in: Scientific American, No. 265-3 (1991) pp. 94–104. http://www.ubiq.com/hypertext/weiser/SciAmDraft3.html.

4 Raffle, Hayes; Ishii, Hiroshi; Tichenor, James: "Super Cilia Skin," in: *Textile*, vol. 2, No. 3, pp. 1–19

5 See e.g.: Nox Architekten with Q.S. Serafijn: "D-Tower: Interaktiver Turm," in: *Archplus – Zeitschrift für Architektur und Städtebau*, No. 172, December 2004, p. 38 ff. and Hubertus Adam: "Der Herzschlag von Doetinchem," in: *Deutsche Bauzeitung* 4/06, 140th year, p. 55 ff.

6 Colomina, Beatriz: "Fort aus der Gegenwart: Das Haus der Zukunft," in: Alexander von Vegesack, Jochen Eisenbrand (eds.): *Open House. Intelligent Living by Design*, Exhibition Catalogue (Weil am Rhein, 2006)

7 Onishi, Norimitsu: "South Korea, Very Wired, Is a Habitat for Robots," in: *The New York Times/Süddeutsche Zeitung*, 10th April 2006, p. 3

8 See Anthony Dunne, Fiona Raby: *Hertzian Tales: Electronic Product. Aesthetic Experience and Critical Design*, London 1999. And: Anthony Dunne, Fiona Raby: *Design Noir: The Secret Life of Electronic Objects*, 2001

9 Interview with Toyo Ito, in: *Thesis. Wissenschaftliche Zeitschrift der Bauhaus-Universität Weimar*, Issue 1/2 1997 (Techno-Fiction. Zur Kritik der technologischen Utopien Vol. 1, 7th International Bauhaus Colloquium in Weimar), p. 9f.

10 Staub, Urs (ed.): *Decosterd & Rahm. Physiological Architecture/ Architecture Physiologique* (Basel, 2002)

11 Rahm, Philippe: "Invisible Architecture," in: *Verb Boogazine*, Conditioning. The Design of New Atmospheres, Effects and Experiences, Barcelona 2005, p. 129 ff. See also Raffael Magrou: "Sensorial Ceremony. Physiological Architectures Here and Elsewhere," in: *Techniques et Architectures* (Imaginaire Scientifique), No. 479, August/September 2005, p. 27 ff.

12 Banham, Reyner: *The Architecture of the Well-Tempered Environment*, London 1984 (first edition 1969)

13 Oswalt, Philipp: "Die Architektur intelligenter Gebäude," in: *Thesis. Wissenschaftliche Zeitschrift der Bauhaus-Universität Weimar*, Issue 1/2 1997 (Techno-Fiction. Zur Kritik der technologischen Utopien Vol. 1, 7th International Bauhaus Colloquium in Weimar), p. 275

VILÉM FLUSSER

THE HOUSE OF THE NET

Houses are made up of a roof, walls with windows and doors, and other parts that are not quite so important. The roof is the important thing: "homeless" and "without a roof over one's head" are synonyms. Roofs are devices to make us subservient: Under them one can cower and hide from one's lord (be he a God or Nature). The German word for roof, *Dach* (like the cognate English word *thatch*), comes from the same root as the Greek *techne*; accordingly, roof-tilers and thatchers are artists. They draw the line between the province of the Laws and the private space of the subservient slave. Under one's roof, the Laws are only valid up to a point. The tree-tops served as a roof for early hominids' nests. We do not think that we ourselves are the ones who project the Laws. We do not need any roofs.

Walls are there for us to defend ourselves against the outside world, not against the world over our heads. The German word for house walls, *Mauern*, like French *murs*, comes from *munire*: to protect oneself. They are munitions. They are made up of two walls: The outside wall turns to face dangerous aliens (lurking on the outside), would-be immigrants; the inside wall turns inwards to the inmates of the house like a jailer responsible for their security. In the case of walls without roofs (e.g. the Berlin Wall and the Great Wall of China), this function becomes clear: The outside is political, the inside wall secretive, and the wall has to protect the secret place of the heart from being visited by evil spirits. Anyone who can't stand secretiveness should go ahead and knock walls down. But even secret-mongers and patriots have to knock holes in walls. Windows and doors. So as to be able to look outside and go out. Before the verb *to show* came to have the meaning "*to display*", it was used in the sense of "*to see*"– i.e. looking out from inside, for

Before the first Internet craze really began, Flusser (1920–1991) published his vision of a type of architecture that was bereft of roofs or walls, which would radically change human existence. Flusser sees the traditional house as a hindrance. It has windows and doors, i.e., holes in its walls. When we leave it to experience something, we lose ourselves. When we return home to rediscover ourselves, we lose the world we wanted to conquer. What Hegel calls an "unhappy consciousness" is for Flusser not least a structural mistake in terms of construction. According to him, a normal house is not open enough for a free life, neither does it provide its inhabitants with security from outside dangers, as once a cave did. It is no longer a unit. The use of many media, which have since increased in number somewhat, has riddled the house with holes for a long time. Although it has windows, the outside world is perceived

which the window provided the device. One looked out from inside without getting wet at the same time. The Greeks called this sort of seeing *theoria*: perception without danger and without experiencing anything. Now it is actually becoming possible to push devices through the window into the outside world to experience things in such a way that one does not run any sort of danger whatsoever. The phenomenological question raised by this is: Are experiments carried out through the window (i.e. in theory) valid? Or does one have to go out the door to experience things?

Doors are holes in the walls for going in and out. One goes out to experience the world, and there one loses oneself, and one returns home in order to find oneself again, and in so doing one loses the world that one set out to conquer. The back-and-forth through the door Hegel calls the "unhappy conciousness". In addition, it can happen that on returning home one finds the door locked. It is true that one has a bunch of keys in one's pocket (or one can decode the secret code), but the secret code may have been re-coded in the meantime. Home and homeland are favourite places for con-tricks. The one is struck outside in the rain, practically in the gutter. Doors are devices that do not provide happiness, nor are they to be trusted.

In addition, the following objection can also be made to windows and doors: One can look in through windows from the outside and climb in through them, and the public sphere can break into a private house through the door. One can of course protect windows from spies and burglars by installing shutters or protect the door from the police by building a drawbridge, but then one lives in fear, stuck within one's four walls. There is not much future in such architecture.

Roofs, walls, windows and doors do not fulfil their function anymore, and this explains why we are beginning to feel homeless. As we can't really go back to tents and caves (even if some people try to), for better or worse we have to design a new type of house.

through the television. Thus there are few possibilities for interaction, information only flows one way. Flusser suggests that we free ourselves of walls, adopt a "curvature of space" as opposed to the house and, starting from the point where those lines stop, create a global interpersonal network. This form of building, the building of alternative worlds, would be a technical challenge not only involving architects and designers.

In fact, we have already started to. Home-as-one's-castle with its roof, walls, windows and doors now only exists in fairy tales. Material and immaterial cables have knocked as many holes in it as in a Swiss cheese: On the roof there is the aerial, telephone wire comes through the wall, the television takes the places of the window, and the door is replaced by the garage with the car. Home-as-one's-castle has become a ruin with the wind of communication blowing through the cracks in the walls. It is a shoddy patchwork job. What is needed is a new type of architecture, a new design.

Designers and architects must no longer think in terms of geography but in terms of topology. Enough of the house as an artificial cave; it is more a warping of the sphere of interpersonal relations. Such a change in the way we think is not easy. The change from thinking in terms of flat surfaces to thinking in terms of the surface of a globe was an achievement for its time. But thinking in terms of topology is made easier by the use of synthesized images of mathematical equations. Thus, for example, one no longer sees the earth as a geographical location in the solar system but as a warp in the gravitational field of the sun. This is what the new house must look like: a warp in the interpersonal sphere by which relationships are "attracted". Such an attractive house would have to assemble these relationships, process them in the form of information, store them and pass them on. A creative house as the nucleus of an interpersonal network.

Such a method of building a house using cables links is full of dangers. For example, the cables can be connected up, not to networks, but to the masses: something "fascistic" rather than "dialogic". Like televisions, not like telephones. In such a worst-case-scenario, houses would be the cornerstones of an unimaginable totalitarianism. Architects and designers must provide a network of reversible cables. This is a task for technology, and designers are up to do it.

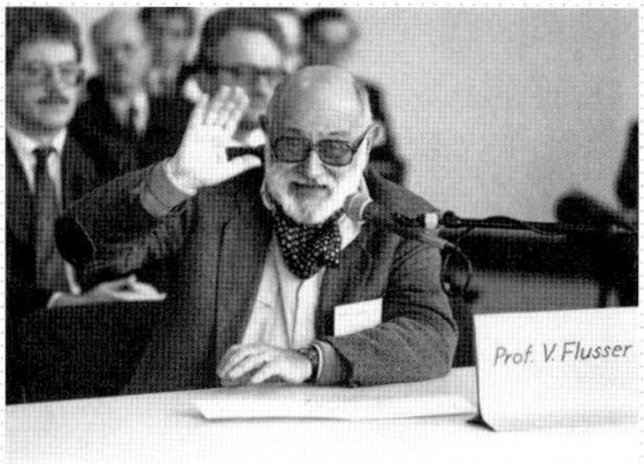

Vilém Flusser at Zeche Zollverein on 5th of March, 1991. On occassion of "Entrepreneur talks" of the company Agiplan he presents his thoughts about the factory of the future, "the place where 'homo faber' becomes 'homo sapiens sapiens' because he has realized that manufacturing means the same thing as learning."

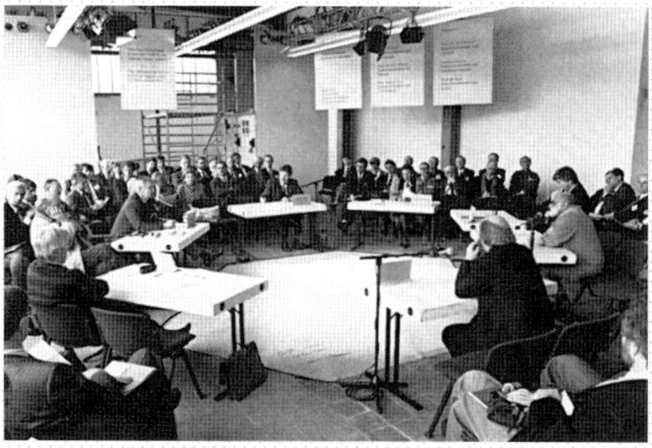

To be sure, such a method of building houses would be a technological revolution that would go far beyond the competence of architecture and design. (This is the case in all technological revolutions, by the way.) Lacking roofs and walls, such architecture standing wide open to the world (i.e. made up entirely of reversible windows and doors) would alter the nature of existence. People would have nowhere to cower any more, nowhere to go to ground or take over. All they would be able to do would be to offer one another their hands. They would no longer be subservient slaves; there would be no Lord over them anymore to hide from or in whom to seek refuge. (Schiller is wrong when he claims that a good father must be *living* above the millions of brothers.) And there would no longer be any Nature threatening them and which they had to dominate. On the other hand, these houses standing open to one another would produce a hitherto unimaginable wealth of projects: Connected up to the network, they would be projectors of alternative worlds accessible to all human beings.

Such a method of building houses would be a dangerous adventure. Less dangerous, however, than hanging on in the ruins of the houses of today. The earthquake that we are witnessing forces us to embark on the adventure. Should it meet with success (and it is not totally out of the question), we would then be able to live again, process noise into information, experience something. If we do not embark on the adventure, we are, for the foreseeable future, damned to huddle between four walls under a roof full of holes in front of our television screens or to drive around in our cars, experiencing nothing.

FRANCESCA FERGUSON

TALKING CITIES THE MICROPOLITICS OF SPACE

Talking Cities is a response to the particular condition of the Zeche Zollverein and confronts a vital issue in architecture and design – that of dealing pro-actively rather than retroactively with the architectural monument. Taking on a central contemporary issue – the regeneration and reactivation of such industrial sites and their surroundings, the heterogeneous collection of architecture and design propositions treat the location as a tactical icon.

My own very first impression of the site was that this was a monument to Bauhaus design, a stylish production machine, rather than a relic and eternal reminder of graft and *industriekultur* – the culture of industry. As a monument it seemed strangely fragile. Designed to last for 30 years and for as long as coal-powered energy remained a profitable enterprise, its naming as a world heritage site ran counter to the radicalism of Fritz Schupp and Martin Kremmer's original vision. Various interventions had created surreal juxtapositions and engendered a sense of place held in suspension: the "works swimming pool" installed amongst the infrastructure of the coking plant, with a defunct fairground wheel rising up behind it, was a collage of machine, ruin, monument and play that distinctly echoed the spirit of that carnevalesque and technocratic blend inherent in the late Cedric Price's never realised "Fun Palace" proposal. This megastructure – a "giant toy", conceived in the 1950s with Joan Littlewood with its carefully researched technological efficiency and flexibility – was intended as a support structure for numerous playful and theatrical interventions. Price's Fun Palace sketches, when aligned with photographs showing ice

Like the World Cultural Heritage Site Zollverein colliery on the underdeveloped northern periphery of Essen, the whole of the fragmented, over-developed Ruhr region is trying to redefine itself. In this context, Francesca Ferguson's *Talking Cities* exhibition enquires into the kinds of possibilities for revitalizing spaces that lost their significance with the decline of industrial production. She is concerned with bringing together the most varied approaches of artists, designers, architects and filmmakers who see potential in the long neglected suburbia and want to focus attention on it. The 1970s saw the first projects and essays on this topic, even architects Venturi and Scott Brown, known for their ground-breaking works including *Learning from Las Vegas*, adopted the theme of life in peripheral zones. The projects in the exhibition address the use of tiny vacant sites in Tokyo and a walk along the disappeared

skaters amongst the giant steel machinery, with the birch trees and pale greenery stealthily engulfing the industrial remains, suddenly mirrored reality.

The paradoxes that become apparent when faced with the ultimate transformation of the Zollverein site are underscored when one explores the urban landscape that surrounds it. In its sheer scale almost a city within the peripheral landscape of Essen, Zollverein is indeed a space to be stepped into if one wants to avoid facing up to the economic realities at its borders – the nondescript social housing estates of Katernberg, one of the poorest areas in Germany, the boarded-up shop fronts with their old-fashioned signage and retro-window displays are almost mocking the stateliness of the industrial monument that was once the economic provider for Essen and Gelsenkirchen.

OMA, commissioned with the masterplan for the conversion of the coal-wash building Zollverein, refer to the site as being a kind of "walled city". By planning the new architecture at the periphery of the site, they preserve the sense of stepping inside another kind of modernity where the commonplace no longer applies.

So the idea of implanting a cultural event with ENTRY2006, one that is to draw attention to new approaches and achievements in architecture and design becomes far more complex and multi-layered as soon as one focuses on the latent contradictions within and around the site itself: an area bedevilled at present by a longer-lasting economic downturn. The preservation of what was once thought as temporary, its cultural renaming as a motor for economic change in the area, and the fragmented landscape, the conurbation that surrounds Zollverein and which is the Ruhr District itself – still in the process of redefining itself beyond its weighty industrial heritage.

Berlin Wall, which can be seen as representing a number of urban conflicts. In a certain sense, according to Ferguson, the diffuse character of today's cities signifies a break with the ideological, utopian visions of the past. The temporary reuse of buildings or permanent intervention in modern architecture make it clear that, in times when public funds are low, we have to concern ourselves with what is left in our cities. Ferguson's collection of contemporary positions on architecture and urban planning thus represents a form of anti-architecture.

Architecture and urban design that engages with ruin and redundancy has been hitherto under-represented within an overall design discourse. And yet it is an increasing necessity, and one which is felt most acutely in the Ruhr Valley, Germany's industrial heartland.

Rem Koolhaas declares this lack of an attitude about the past as a weakness in modern architecture. Cedric Price – described by Koolhaas as "the architect's anti-architect" for propounding values that question the very need of built architecture, and upholding the temporary, the subversive and the playful in the face of any form of architectural dogma, was one of the first in his profession to think profoundly about the past. Whilst the Fun Palace proposal was theatrical, in tune with the communal – even ludic – spirit of the 1960s, His famous proposal in 1969 for turning a Potteries Factory in Staffordshire into a *think-belt* – a basis for education and change – confronted his own very real concern for the industrial heartland of his youth.

The Thinkbelt proposal could well be transplanted to the Ruhr of the present day, where in sites like Zollverein the change is already taking place. Koolhaas is adamant in his admiration of the Price's vision: "Its sensual feature is its concern with the reuse of an abandoned part of infrastructure: the railway... a way of re-inhabiting an existing condition that was no longer functional. In this way the Potteries *think-belt* dealt with a dysfunctional abundance, the abject, the ruin. Price found a way to inscribe a new programme... He transformed a previous programme into a new programme by taking the derelict and using it as a basis for his new work."

The conceptual link to *Talking Cities* thus becomes a foundation for architectural and design practice that confronts the derelict, the abject; urban spaces and built structures that have lost their former meaning and significance. The questions then asked by

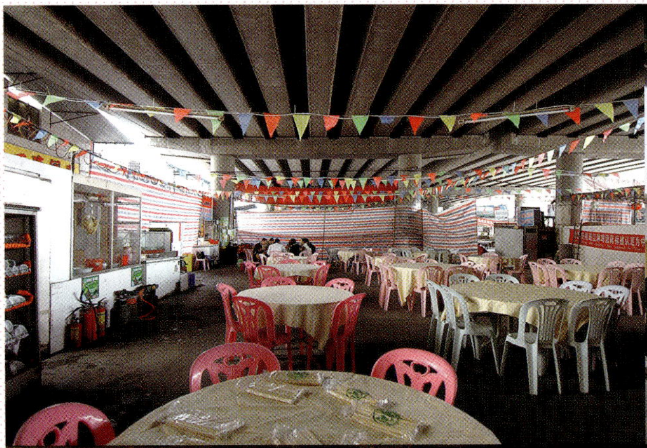

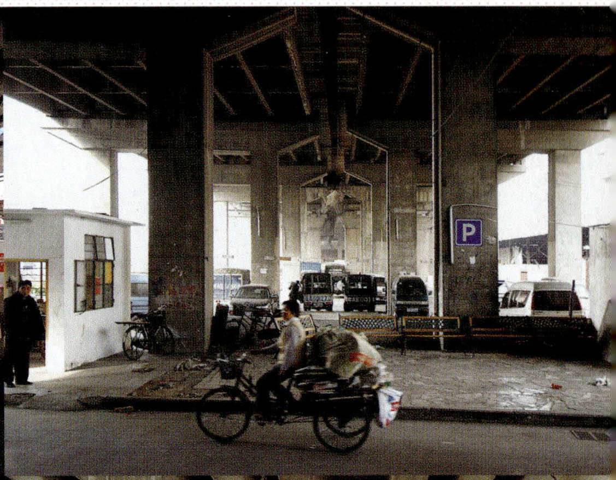

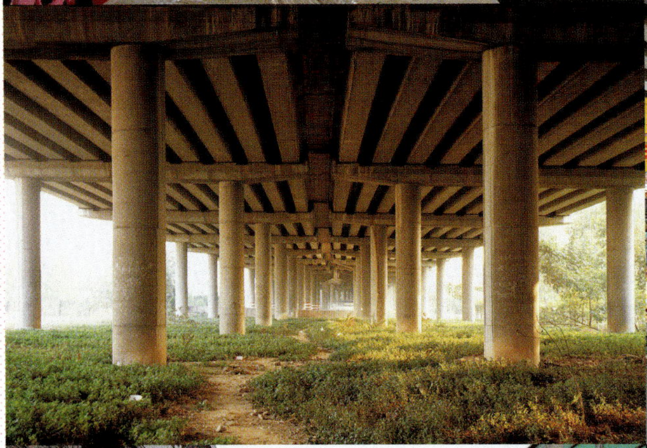

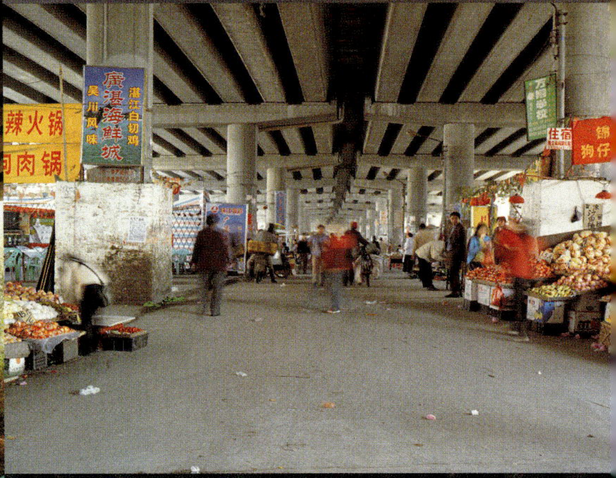

辣火锅
肉锅
廣湛海鮮城
吴川风味
湛江白切鸡
住宿
狗仔

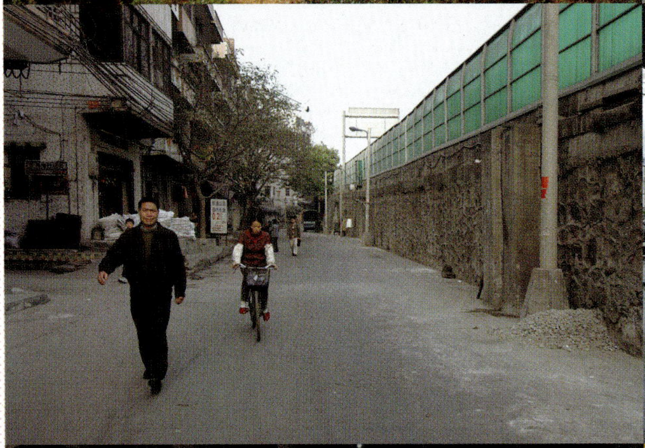

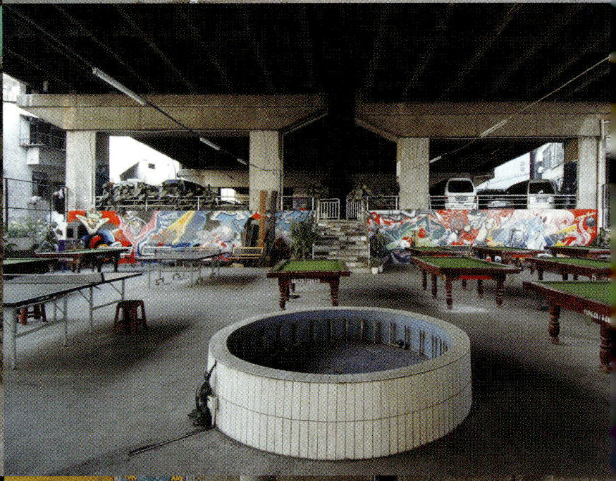

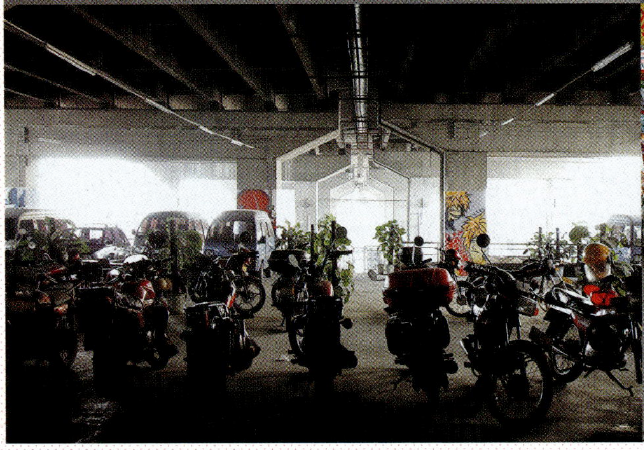

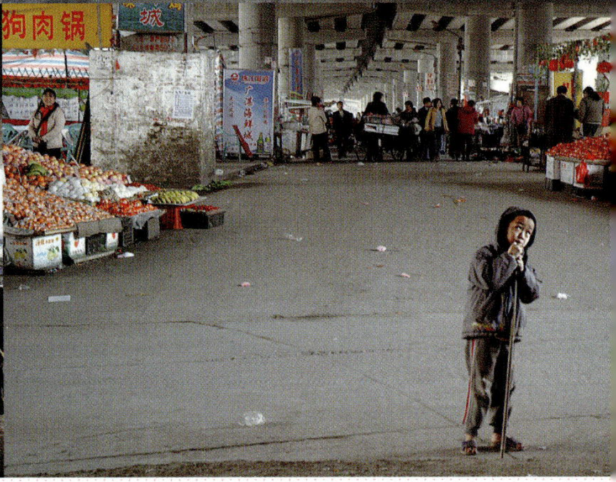

狗肉锅
城

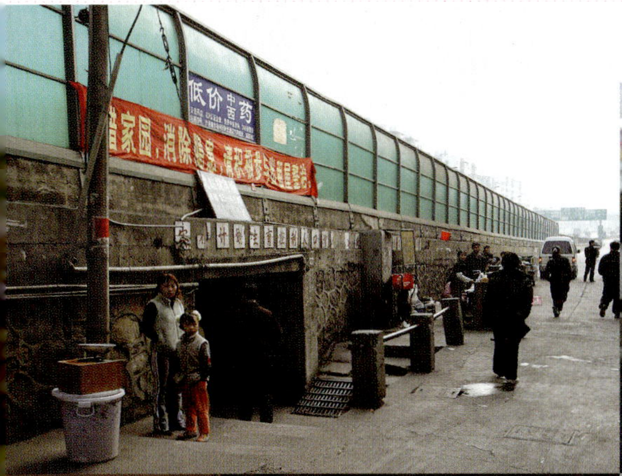

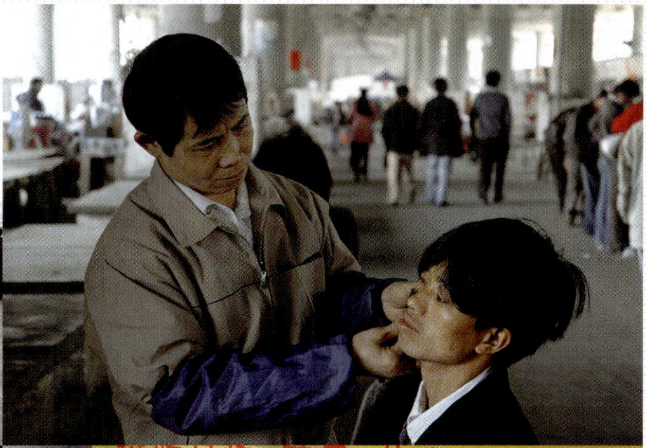

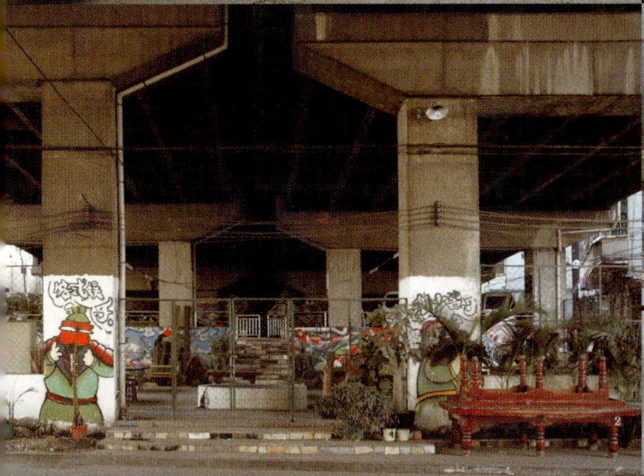

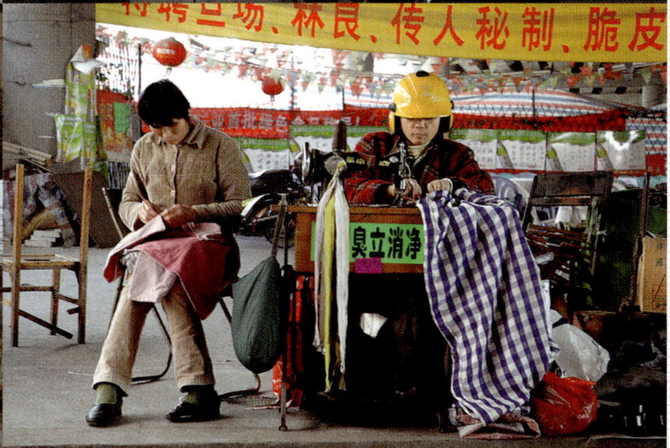
传人秘制、脆皮
臭立消净

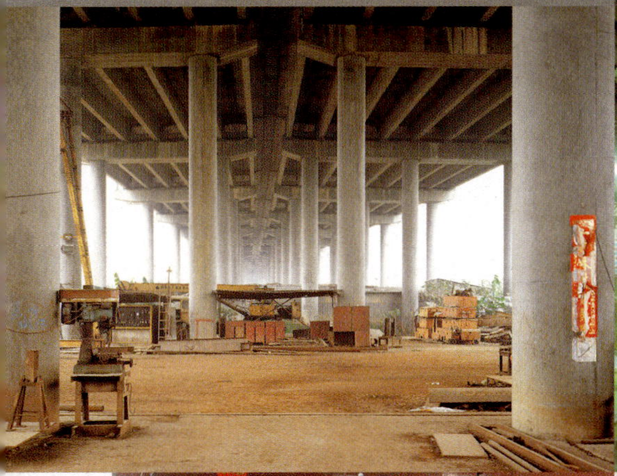

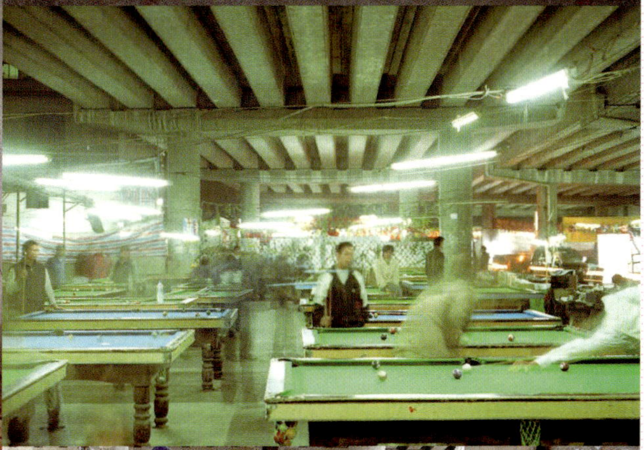

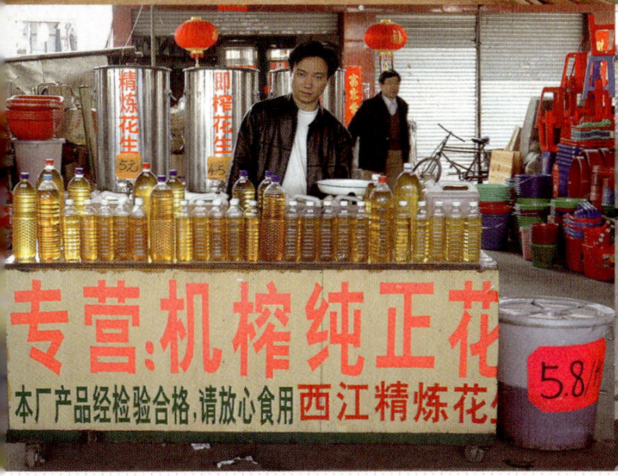
精炼花生
即榨花生
专营:机榨纯正花
本厂产品经检验合格,请放心食用 西江精炼花
5.8

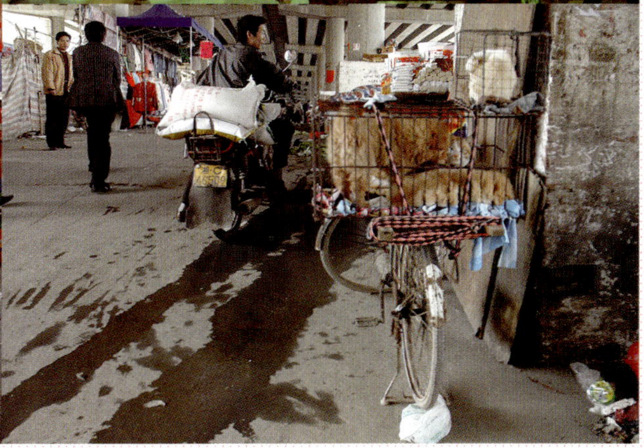

‹
Photo series: "Underneath", Gutierrez+Portefaix, Map Office, Guangzhou, 2005. New usages underneath the city expressway, where migrants from rural communities found a new living space.

"Architecture should once again be seen as a cultural signifier, as part of a dynamic condition."
--- Jay Merrick, architectural critic, The Independent, UK ---

the protagonists and the specific urban and suburban conditions within which they operate are: how can architecture and design practice re-engage, reposition, re-ignite spaces that are undervalued, in transition, or at the margins of perception? This is a conceptual focus that takes inspiration from both politically, socially and economically challenging contexts – and necessitates resourceful, and at times radical design propositions.

Peter Lewis draws attention to Price's definition of cities as concentrates. If so, *Talking Cities* is a concentrate: of propositions, designs, and urban explorations. The dialogic inserts itself within the mega-structure and proposes action on another scale. To be viewed perhaps as a collective enterprise, the discursive joining together of such current practice underscores the political focus of individual propositions. Within the space of preservation and heritage, micro-topic interventions and itineraries create an overlay or plug-in that is both useful and affordable. The breakdown of industrial-production requires not spectacle and awe but action which draws upon the potential of individuation – the possibility of inscribing new realities upon outlived monuments, and transplanting another human/pragmatic scale within the monumental. Hans Ulrich Obrist saw Cedric Price's agenda as creating a "laboratory of delight, a university of the streets" and it is with both the experimental and the educational in mind that *Talking Cities* enters the coal washery as a community of practitioners delighting in the state of transition.

Doubt, delight and change – Possibly the main mantra for Cedric Price's approach to the discipline of architecture becomes the basis for the collective statement within *Talking Cities* amongst artists, designers, architects and filmmakers – one which takes inspiration from the current urban spaces of fragmentation, divisiveness, and marginalisation. "The architect ought to be an improviser rather than someone who wants to rule" –

"— so that is another rule for the whole nature of architecture; it must actually create new appetites, new hungers—not solve problems; architecture is too slow to solve problems."
--- Cedric Price in conversation with Hans Ulrich Obrist (RE:CP) ---

declared the architect Peter Cook (co-founder of the group Archigram), and it is the design of customisation, of tuning and plugging in that unites the protagonists in a conversation on designing for contemporary urban landscapes. Far removed from the grandiose mega-visions of the modern movement, designers are increasingly acknowledging the need to review what such eras have left behind and to extract what is adaptable, "tunable", re-usable from that which is declared to be redundant.

PLUG-IN – THE INFORMAL CITY

Even the exhibition design by the architect Thilo Fuchs foregrounds this informal, low-cost stance in design with a flexible structure. A wooden grid system, literally wrapped around the existing steel columns (whose heritage preservation disallows the smallest nail or incision) and mimicking the rust-red infrastructure that distinguishes the Zollverein site, supports a dense, heterogeneous collection of statements and projects. The grid system allows for small-scale adaptations and at the same time upholds the practice of re-use, of bricolage, and of singular design statements composed of makeshift elements. The dialogical aspect of *Talking Cities* as an evolving conversation amongst the protagonists, binds the individual contributions through graphic strips bearing statements and reflections upon the themes. To cite Cedric Price: "Dialogue might be the only excuse for architecture... Creating a continuous dialogue with each other is very interesting. It might me the only reason for architecture – that's the point!" (RE:CP – Hans-Ulrich Obrist).

DISPERSED URBANITY

Talking Cities takes its inspiration, and develops its central themes, not only from the Zollverein site itself and the conceptual links that it suggests, but also from the wider

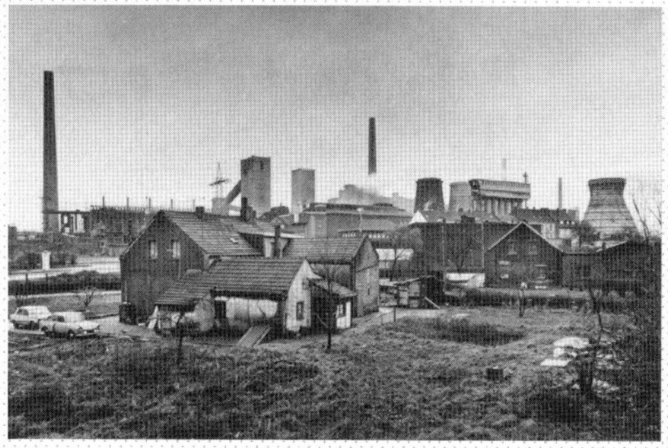

Osterfeld, Oberhausen: Osterfeld mine, 1976 (on the l.), Altenessen, Essen: Hohendahlstrasse, 1977 (below), pictures from the series "Stadt- und Industrielandschaft im Ruhrgebiet" made by photographer Joachim Schumacher between 1973 and 1987.

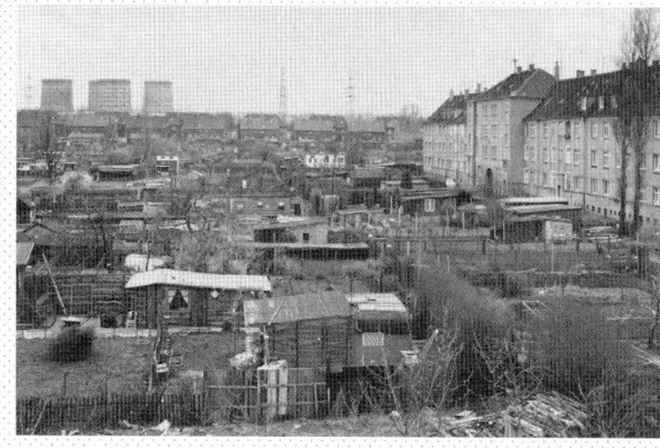

setting of the surrounding Ruhr Valley itself: A dense conurbation of towns and cities, the borders of which are – to an outsider – not self-evident. Economically the area is coping with the long and fraught process of re-identification with a new reality, one that requires the wholesale transformation of industrial sites, either demolishing them altogether, allowing nature to reclaim the territory, or injecting them with other social and cultural uses – as is the case with Zollverein. The agglomeration, the relentless suburbanisation of the contemporary city is a main point of departure for current practice – and a necessary area of investigation for today's architects and urban planners.

These are fragmented urbanised and marginal landscapes where industrial monument, housing estates, suburban allotments, business districts and shopping malls coalesce. They consitute a contemporary condition to which most architects and planners have kept their distance – preferring instead to celebrate density, the metropolitan. However, the economic realities of such territories can no longer be denied: the Centro shopping mall in Oberhausen, decried by architects as a blot on the landscape and flanked by an aquarium and the vast Gasometer – the largest of its kind in Europe – in a seemingly random post-industrial assemblage of monument, spectacle and consumption, in fact boasts up to 80,000 visitors a day.

Susanne Hauser refers to such territories as "an-aesthetic", meaning that we simply lack the language and the tools to perceive these areas; they lack classical imagery. MVRDV described the Ruhr District in their polemical and futuristic study *Rhein-Ruhr-City* as a "dispersed and low-contrast periphery. Between Unna and Duisburg, Gladbeck and Wupppertal, there is hardly any big-city urbanity. It consists rather of suburbs, green spaces and old industrial surfaces. The Ruhr District resembles a conglomeration of in-

Eisbahn coking plant at Zollverein Essen, the "Zollverein Coal Mine Industrial Complex" was included in the UNESCO World Heritage list in December 2001 – from the end of December 2001 until mid-February 2002 you could skate along the sets of coking oven.

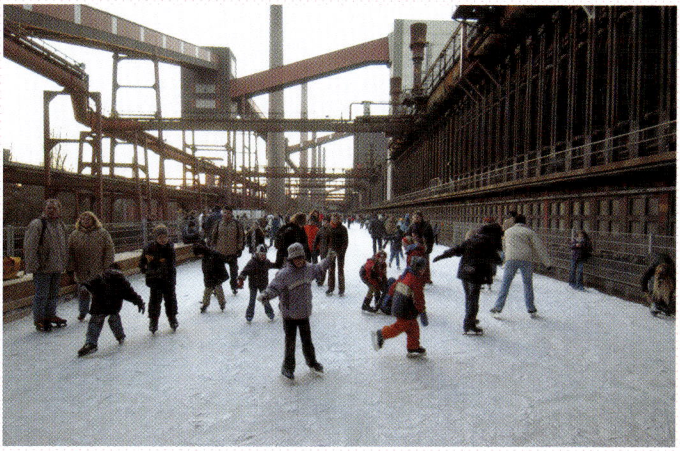

between areas, where many of the cities resemble each other. Here, one does not find real borders between city and country."

If this is the current condition of cities beyond approximately five per cent of the area that consitute their historical core, then the photographs taken by Joachim Schumacher of the Ruhr Valley at the height of the industrial period presaged this thoroughly modern development: The extreme pace of industrial development created territories where subsistence economies flourished adjacent to industrial mega-structures: giant chimneys and water towers are flanked by small allotments and urban gardens, which give way to a few rows of one-family houses.

Stitched together with an excellent infrastructure and mainly perceived from the point of view of driver or passenger, the Ruhr conurbation of the present reveals to those who explore its fringes a plethora of interstitial and leftover places. This is a landscape that requires a pioneering insight, a new reading. The artist Boris Sieverts and his partners – urban planners and architects – initiators of the project "Land for Free" see the rough and abandoned scapes of the Ruhr as denoting another kind of beauty and freeing up opportunity for land occupation with a whole new aura. "The bricolage character of the contemporary city is what interests us". By proposing a radical deregulation of property on the margins of industrial landscapes, in negotiation with the local authorities, pockets of land could – according to their proposal – be handed out to bidders based upon the originality of proposed designs and uses. By seeing the liminal spaces in summation as 'the other city' they propose a kind of liberation from the complicated networked landscape of the Ruhr. "Every good idea is blocked by static property relations", declares "Land for Free". So why not see the leftover landscapes of obsolescence as spaces of opportunity for individual occupation? In a wry take on contemporary globalisation and the trans-cultural habits of the contemporary urbanite, one supporter of the project declared the terrain of a typical brown field site, with black slag and bright green of birch trees, as a space of desire: "Perhaps these areas can be hyped in some way: mansard flat in Paris; houseboat in Amsterdam; loft in New York and a little birch wood on a coal infill – perhaps even a hillside plot on the spoil heaps of the Ruhr area." Stefanie Bremer, "Land For Free"

The architect Simon Conder took on similarly pioneering territory with the rubber house – effectively a customised shed, clad completely in black rubber – in the edge landscape of Dungeness – "a place of strange alienation and rough beauty" dominated by a huge 1950s power station and the largest shingle beach in Europe. The fetishistic and radical design – albeit on a small scale and with a limited budget – by the architect is a statement of freedom gleefully described as a place of architectural lawlessness. Such designs represent an approach to marginal territory where conventional planning regulations no longer apply, and in which design practice is also liberated from such constraints.

Any legitimate form of architectural intervention depends on such subjective itineraries of the "edge" landscapes of the present. In fact, at a time when the edges of cities and towns seem out of reach for architects and urban designers, determined largely by economic rather than aesthetic concerns with their networks of supply, consumption and storage, the architect and planner needs to develop entirely new tools for mapping such territory. Bernd Kniess and Leonard Lagos declare the architectural profession trapped in its own expertise – with outmoded definitions and images. Their strategy is one of "reading" and tracing what goes beyond the conventional definitions of the city through the eyes of a younger generation of "users". The active, experimental and un-fettered use of space within the Ruhr conurbation by young people, by migrants, and in which telecommunications technology plays an integral part, is central to their understanding of the Ruhr conurbation as a highly relevant territory, and one that can shed greater light on new urban conditions.

> "The big topic of today, and of the next 20 years,
> will be peripheries.
> How you can transform peripheries into a town.
> The peripheries are the city that will be.
> Or not. Or will never be."
> *Renzo Piano*

Mapping and charting territory that eludes classical definitions thus becomes a primary pursuit of architects coping with the fact that opportunities to build and design at the saturated and thoroughly planned historical core of cities are eluding their grasp.

The spaces with the greatest potential for redefinition and redevelopment form part of our "suburbanised" reality. The architectural critic Valéry Didelon – teaching "suburbanism" in the Architectural school of Versailles points out that faced with these realities, "architects must let go of grand notions" – forget "urbanizing suburbia", but "embrace suburbia, incorporating its mode of propagation, accommodating its aesthetic and participating in its expansion". Like the architects Venturi, Scott Brown and Izenour, who shook up the architectural profession in the 1960's with their reading of vernacular architectural in *Learning from Las Vegas*, it was by venturing outside the city that they managed to strengthen their art.

As soon as one embraces and defines the territories and landscapes that indicate our contemporary condition, the opportunities for a re-positioning of design culture within the dispersed, transitory urban reality that makes up most of our surroundings becomes real. IaN+ an architectural practice based in Rome placed an investigation of the vast ring road surrounding Rome – the GRA (Grande Raccordo Anulare) – at the centre of their research. The negative space within which most Romans spend time in endless traffic jams is a space of transit that divides the ancient city from the ever-increasing sprawl. The GRA as a filter between two realities becomes, once traced, explored and mapped (and to this end photography plays a vital role), a psychological gateway between "belonging" to the true, intact and historical city, and facing the realities that are inexorably spreading beyond it. IaN+ defines a design strategy celebrating urban voids and emptiness, as a way of counterbalancing the relentless spread of the city which develops without a direct relationship to the landscape: As an urban design proposal resulting from such visual mapping strategies, empty spaces would literally be walled in by buildings and these voids act as a built limit upon the relentless expansion of the cityscape. Such practice is a radical reversal of the mundane, the unplanned, to manifest an alternative urban design strategy.

An innate fear of the unplanned and empty urban spaces – the derelict and the void – seems so deeply anchored in the minds of contemporary – western – culture, and perhaps nowhere more so than in Berlin. Here the border and the wall that separated the city and two ideological worlds were symbolic of a collective trauma that had to be erased as rapidly as possible by both the local political interests, and in the minds of the citizens. It took several years before "monuments" to the wall, in a post-hoc reconstruction for the city's tourists, were erected on sites where the incision in the city's fabric was most painfully felt. The politics of *détournement*, of *dérive*, in order to bring to the fore this now largely overgrown and often invisible border around the city, was carried out in a "wall-walking" exercise initiated by the Italian architects studio.eu and Stalker in order to revive a consciousness of urban transformation here – both physical and psychological. It is another kind of seeing – one from the point of view of the stranger and untram-

Nuno Cera: "Ultra Ruhr", Ruhr District, 2006, commissioned by Talking Cities/urban drift productions, film stills

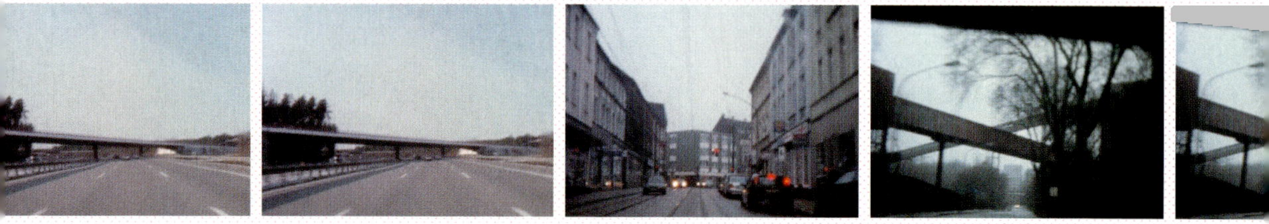

meled by the collective historical trauma of the city's recent history, that predetermines any kind of planning to imbue such a border space with new meaning.

> "We see it [the space of the Berlin Wall]
> as a concentrated manifestation of many of the sources
> of urban conflict that are prevalent everywhere." –
> *studio.eu*

THE FRAGMENTED GAZE FOR THE "EDGE CONDITION"

> "All I need is a brief glimpse, an opening in the midst of an incongruous landscape, a glint of light in the fog, the dialogue of two passers-by meeting in a crowd, and I think that, setting out from there, I will put together piece by piece, the perfect city, made of fragments mixed with the rest, of instants separated by intervals... discontinuous in time and space, now scattered, now more condensed." –
> *Italo Calvino: Invisible Cities*

The artist and filmmaker Nuno Cera sent this quote as an indication of his filmic perspective on the Ruhr District. His photographic investigation allows for a melancholic and scattered reading of the landscape – and his filmic method of aligning thousands of 8mm film stills, an extreme slicing of perspective followed by the re-forging of a whole made of a myriad of visual fragments and impressions, frames an instant subjective "discovery" of a territory; an intuitive perception of the landscapes of the edge condition.

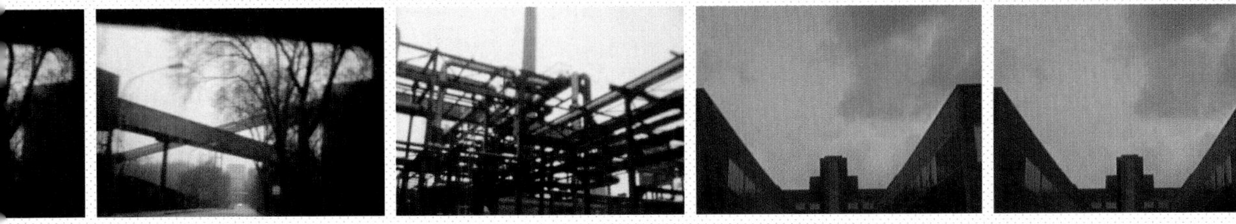

Photography and film is one of the most powerful tools for charting and documenting dispersed and often randomely planned urban territories and determining the way in which such territories are activated and used. However in the spaces of dispersed urbanity that determine a reality at the margins of the contemporary city the iconographic approach to the image to which we have become accustomed can no longer be applied.

> "The city as we know it seems to be dissolving and
> is being replaced by something for which we lack concepts and images."
> *Kai Vöckler*

The Dutch photographer and artist Bas Princen is one of the contemporary photographers casting aside usual modes of representation that are no longer useful: within the Dutch landscape which has been over-planned and programmed with new recreational landscapes – providing spaces of leisure; safari parks, motorcross circuits, rowing canals, he observes that it is far more likely to be equipment, "gear" and products that determine how any piece of unplanned and undeveloped land will be used. The digital era with its internet communities and ever-decreasing scale of technological equipment defines for Princen "distinct places, biotopes occupied in a specific way by certain individuals or specific groups... The gear makes you aware of landscape features and conditions you wouldn't even notice otherwise." His images of fishing, birdwatching, motorcross activities require close observation before people's minimal occupations of otherwise dull flatlands, fields and dunes become evident.

That a certain perception of spaces, a peripheral, fragmented gaze, necessarily precedes any form of intervention or design is unavoidable in the "an-aesthetic" spaces of our contemporary cities. That photography and film are rarely reference points lies perhaps in the emotional force of their subjectivity. Within *Talking Cities* the protagonists are embellishing the more scientific tools of urban research by allowing for this subjective gaze with which to comprehend and activate hitherto underrated areas for their own design practice. In a sense the dispersed nature of today's cities ruptures the large-scale, ideologically driven utopian visions of the past and, in revisiting those often failed planning visions of past decades, calls into question the hitherto accepted modes of analysis and design.

RE-THINKING MODERNISM'S MEGA-VISIONS

In a typically provocative statement, Cedric Price threw all comfort in the certainties of the architectural profession overboard when, in conversation with Hans Ulrich Obrist in September 2000 he attacked the profession's ideological imperative "It's a way of imposing order or establishing a belief, and that is the cause of religion to some extent. Architecture doesn't need this role anymore; it doesn't need mental imperialism; it's too slow, too heavy, and anyhow, I as an architect don't want to be involved in creating law and order through fear and misery."

In one fell swoop he rejects the mega-visions of the modernists who, after World War Two, experimented to extremes with the peripheries of the cities. Perhaps what drove his statement is the present-day reality that planners are all too often left with the failures of such large-scale planning. As the architectural historian Simon Sadler points out: "Architecture since the seventies atones the sins of a couple of generations long since retired or dead". Valéry Didelon in his essay "Architects on the Edge" links this to the postmodern 'return to the city' that motivated a generation in the 1970s and 1980s, when "(architects) were so unwilling to acknowledge the spread of progress to the suburbs that it escaped them altogether. Instead, they seemed to link their fate to that of the historic city, and embellishing and improving city centres and the surrounding neighbourhoods became the only legitimate focus of their profession."

For many of the architects participating in *Talking Cities* the remains of utopian planning visions are all too evident – such that they can no longer be ignored. The modern movement has left behind mega-structures such as France's vast housing projects, whose isolationism and mono-functionalism were blamed in part for the eruption of riots throughout Paris in the summer of 2005. As Rem Koolhaas put it "All our calculations are categorically superseded by events. And that in itself is generating a kind of overall modesty which gives the current situation a pretty non-utopian quality."

Hence a fundamental shift in scale amongst the practitioners constituting the collective statement that is *Talking Cities*. Taking as read the existing social and economic realities within their own urban context – defunct social housing projects, brownfield sites, undervalued public spaces – the interventions and temporary strategies implemented are micro-political acts. Individuation; the freedom to adapt existing structures on a smaller scale and for specific, singular uses; to implant architecture as modules, is seen as a guarantor of individual liberties.

The architect Luc Deleu exemplifies this approach in his own revaluation of Corbusier's Chandigarh. The fresh view of Corbusier's vast modernist development takes a step beyond the wholesale critique of the radical urbanist – such as was unequivocally aired by Peter Davey in the *Architectural Review*. "The Capitol is flawed by the absurdity of gargantuan scale, daft indifference to function, and total ignorance of construction (Corb never seems to have heard of climate, or even of a drip, hence tremendous dilapidation). Yet the spaces inside the Capitol buildings have generosity, and the notion of a dignified *res publica* in a democratic society. Sadly, political developments have severed the buildings from the community.

Now clearly the city needs to grow, and perhaps its best chance is to make a new area of planned expansion beyond the Capitol, so making it the centre of a city which has made us all think – and try to learn." Peter Davey; "Celebrating Chandigarh", *Architectural Review*, February 1999

Luc Deleu revisits Chandigarh and reads a wholly different agenda into Corbusier's masterplan: "I clearly understood his intuitive search for an order that could integrate ad hoc life and human chaos." Deleu had photographed the human-scale adaptations of Corbusier's buildings – to accommodate snack kiosks at the base of exterior balustrades, for ecample. "The work of Le Corbusier represents a unique design attitude focussed directly on the human being. His strictly formal, urban prototype, 'Une ville contemporaine pour 3,000,000 habitants', evolved via 'plan Obus' into Chandigarh creating an infrastructure where each family can organise its dwelling and environment in almost complete freedom within a positive and stimulating larger order. One can tell he had a collage city in mind."

RESTORATIVE DESIGN – ADAPTIVE ARCHITECTURE

This extolling of the virtues of large-scale architectural infrastructure is indicative of a wider tendency to a pragmatism in architectural and urban design practice that is perhaps generated from the reality that a vast number of existing modernist buildings need to be fundamentally reassessed, reprogrammed, and often saved from extinction. Each micro-adaptation of the large-scale becomes, to the architect as re-investigator of failed modernist visions, a political statement. "Developing the large scale from the small scale ensures a differentiated result because it takes into account peculiarities, exceptions and individuality: it is an essential guarantee for the conservation of individual freedom", so Luc Deleu.

Such adaptive planning strategies are necessary acts where the megavisions of urban planning in the seventies have outlived themselves. Contemporary architects often have little choice but to draw new potential from obsolete programmes. The State Theatre in Darmstadt is a case in point. The German architect Arno Lederer, (of Lederer+ Ragnarsdóttir+Oei) was faced with an outmoded structure of a drive-in theatre – built as a paragon of American-style prosperity. The building of exposed concrete, marble cladding and corrugated iron was a modernist achievement yet – "the futurism of that time returns now as a grand dame, unsteady on her feet". With a glowing module clad in glass tiles, swathes of colour inserted into the bare functional ground floor, and an open stage above the parking garage, the building's facelift exemplifies such restorative architecture, a prime example of a niche which contemporary architects can occupy with ingenuity; effectively prosthetic design for an amputated modernism. Arno Lederer compares such practice to the minimal additions needed to revive an old suit jacket – the metaphor is also a comment on the modest resources available to the architect for such architectural adaptations.

In a pertinent manifestation of popular discontent, the prime-time television series *Demolition*, produced for Channel 4 in the United Kingdom, staged a poll that voted Cumbernauld Town shopping centre in Scotland, once the centrepiece of the 1960s visionary new town of Cumbernauld as Britain's worst building: "large sections are empty but it is home to a smattering of offices, a library and a low-rent shopping centre connected by a bewildering array of ramps and walkways...'hundreds of people voted to demolish this giant superstructure which was built in the 1960s as part of the government's ambitious plans to deal with the problems of urban overcrowding'. (Channel 4) In spite of the wholesale rejection of the rundown site, there was no funding to carry out the demolition of the site. Instead Cumbernauld's town planning department has instead commissioned a group of architects to regenerate the site in a massive facelift... This exposes a typical weakness of under-funded local authorities when faced with the radical town planning of the past, and one that necessitates a pragmatic design approach towards existing – unloved – infrastructures.

TRANSFORMATION

> "Today, after several decades beyond the ideologies of the 1960s architects carry a new perception which is once again more attentive to the movements of the avant-garde and their modes of interventions. The protagonists, architects and designers of the radical movement, in Britain, in Austria, in Italy, produced instruments for contesting realities destined for the citizen, and have opened the urban space as one for utopias and experimentation."
> *Petetin & Grégoire, Paris*

The architects Petetin & Grégoire, and Lacaton Vassal, are engaged in drawing attention to precisely those potentials for adaptation and transformation that could halt the governmental policy in France of a tabula rasa of numerous social housing projects developed in the 1960s. At a time when social housing is in great demand throughout the country, Lacaton Vassal propose opening out, and freeing up the existing architecture by enlarging windows, adding balconies. Petetin and Grégoire intervene in the under-used public spaces within the housing developments with "para-cities" – minimal shells and movable modules made of light materials become tools that can allow for cooperations between the decision makers and the citizens.

The adaptations of mono-functional structures could be regarded as manifestations of Foucault's Heterotopias, or spaces layered with new meanings, of re-invention. Indeed, such short-term, tactical interventions as are the case with para-cities are inspired by non-places beyond the real and the planned.

The architect can also operate as a documentor of such individual re-programming of space – as a micro-political act, generating a diversification of use within the mono-functional infrastructure. The Roman architects Stalker, with Osservatoriao Nomade, applied this practice to Corviale on the south-western perimeter of Rome. This 958m-long building planned by Mario Fiorentino in 1974 is a thorn in the eye of architects and local politics alike, many of whom have called for its demolition. In a trans-disciplinary venture, with photographers like Armin Linke documenting traces of the history of the building, all the ways in which the residents had adapted the building to their needs were used as evidence. "We tried to individuate the instruments that could revitalize a deteriorated relationship between public, administrators, architects and residents." Even the so-called "free floor", the fourth floor conceived as an internal road with shops and services that never arrived, was upheld as an example of spatial occupation that holds potential for further development. Occupied now almost entirely by self-built apartments organised around communal spaces, the fourth floor was squatted, and upon intervention by the architects the illegal practice has subsequently been accepted and legalised by public administrators.

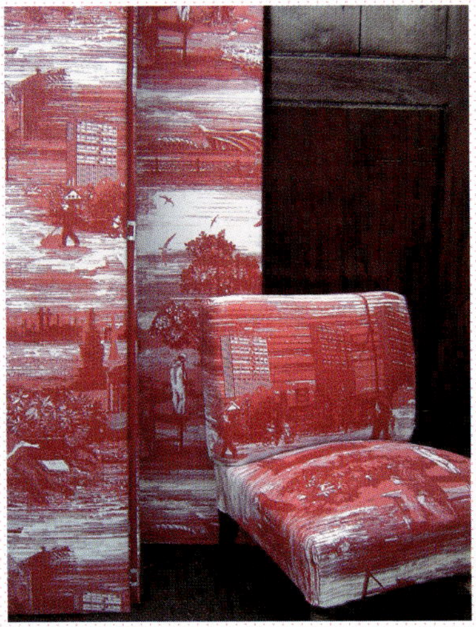

Timorous Beasties, "Glasgow Toile," wall screen and chair covers with stretch covers. Uncompromising, contemporary urban motifs are displayed on traditional fabrics and wallpaper. Contemporary political and social themes are presented in a pastiche context of Romantic georgic from the 19th century.

The architects involved in this process of re-valuation, re-assessing what has been declared as defunct modernism are casting aside their own planning and design preconceptions and drawing upon the self-built, the illegally occupied and adapted as both a political act of diversifying the spatial potentials of existing architecture, and as a source of inspiration for future design. It is the collective imaginary that is activated here, with a fusion of outlived modernisms and enlivening adaptations – no wholesale dismissal of architectural eras predicated upon drastic change, but instead a wry modesty in engaging with their leftovers.

As the relics of similarly large-scale, prefab socialist housing projects in satellite towns throughout Eastern Germany and Eastern Europe are abandoned, with up to 40 per cent of the population taking advantage of open borders and better employment opportunities, a significant design practice of collecting and preserving has emerged to counteract the culture of forgetting. Memory card games made up of numerous architectural details and ornaments of the socialist era in architecture (Edition Faltplatte), retro-chic wallpaper prints and patterns reminding Berliners of the seventies icon to socialist culture – the Palast der Republik, palace of the people's republic, are design transformations and counter-strategies. On the one hand a concerted effort by designers to preserve a sense of the socialist aesthetic, on the other a cynical comment on disadvantaged social housing projects on the peripheries that are all too often left to their own devices: the designers Timorous Beasties have embedded the ugly realities of Glasgow's tenement quarters in their design for *Glasgow Toile*. Both design approaches bring the marginalised to the forefront through subtle re-inventions.

Peter Lewis comments the "space of invention" that emerges from residual cultures, a main focus of *Talking Cities* as "a sort of mixed, joint experience in which the real sites within a culture are simultaneously represented, contested and inverted, as in an anamorphic [distorted] lens or mirror." (Peter Lewis: "The evacuated field", in: *Talking Cities. The micropolitics of space/Die Mikropolitik des urbanen Raums)*

ADAPTATION AND THE PRAGMATIC PLUG-IN

Kees Christiaanse, an architect now instrumental in incorporating the strategies of temporary use, intervention and occupation into his urban masterplans, declared his processual approach to urban design, which allows for temporary uses and occupations as an integral part of his pluralistic urban planning concepts to be informed by his involvement 1970s culture of communal living and squatting. It is those who, like him, were once illegally occupying low-cost housing at the centre of Amsterdam for example, are now the political decision makers creating the conditions for greater participation of the citizen in design processes.

Perhaps also the uncomfortable ambivalences of the neo-liberal climate of the nineties in which transitory, temporary and mobile architecture employing the latest lightweight technologies was admired for the style and gloss it might add to the urban fabric – and often preceded gentrification – has given way to a more realistic planning approach; one generated by recession economy. The lightweight mobility, fuelled with the positivism that accompanied the new-media economy, was far from being an affordable resource available to the many. Since the new-media bubble has burst, the mediation of planning processes has advanced to accompany this recession economy; where the incorporation of existing spatial occupations and adaptations, is an acknowledgement of limited budgets and a more modest approach; one that concentrates upon creating conditions within existing structures.

This is a far cry from the cartoonesque provocations by Superstudio, Archigram and the utopian mix of plug-in and play that motivated a generation of designers in the 1960s. Current practice and the next generation of urban designers reveal a far greater necessity for prioritizing considerations of political economy in design. A modesty of means combined with a practical "win-win" symbiosis within the regulated world of the property developer induced Floris Schiferli and Tobias Lehmann to develop Nohotel. Proposing a modular system that employs scaffolding to support lightweight components that made up a living/sleeping unit, Nohotel uses existing buildings as hosts. They configure a living space with soft and easily transportable tools. The designers have developed an itinerant architecture that makes squatting spaces an economically viable solution for both host building and parasite structure. This design statement is one of both economy and good behaviour – replete with the kind of sales talk reminiscent of the startup: "the new living standard for temporary housing with all the exclusive qualities thereof".

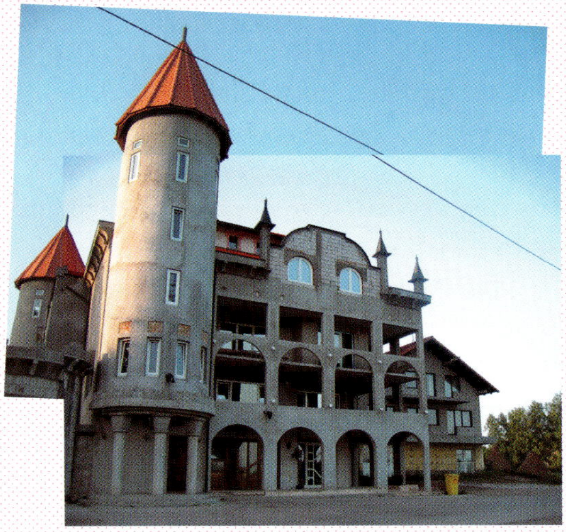

Incomplete wedding palace in Tuzla, Bosnia, 2005. Halls for public events are built using a blend of Christian and Islamic religious architecture.

BALKANISATION IS ARCHITECTURE – THE INFORMAL VERSUS FORMAL ARCHITECTURE

The breakdown of economies and political ideologies also forms the backdrop for both this pragmatism and for a wholly different reading of city spaces. Foucault's Heterotopias – spaces of self-invention – are manifest within informal architecture, the self-built and the self-organised. The political fragmentation is traceable within the architecture of Eastern European cities, where Srdjan Jovanovic Weiss, an architect investigating the plethora of "beautiful ugly" styles in illegally built structures and architecture has adopted the term "Balkanisation". For architecture: If you lose yourself in Europe, where fragmentation is a source of fear, then find yourself in the experience of the Balkans, where fragmentation is normal. Charting Balkanisation in architecture, he draws inspiration from Belgrade's extensive roof top additions, "split-level cute mushroom houses that are a mix of high tech and neo-Byzantine styles: part new Orthodox shrine and part glitzy turbo architecture".

The artist Marjetica Potrã also focuses upon the positive connotations of political economies in dramatic transition by documenting informal architecture throughout Eastern Europe in *Europe Lost and Found* and declares the need to re-asses what is beautiful: Today it is the layering of old and new in the city that determines its particular qualities, even beauty... The big modernist projects in the West Balkans are slowly rotting away, and people are trying to consolidate their territories now. I think that actually the informal city developing in this way is much more optimistic."

In a notorious manifestation of this optimism, in Tirana the mayor Edi Rama has accepted the illegally built structures and merged them with new developments by having the whole city painted in bright colours.

The use and adaptation of basic infrastructure and the current celebrations of the informal, the self-built, denies the need for an architectural language of style, as though the favela, the kiosk, the bricolage of personalised styles and micro-realities, with the tenacity of their survivalism, mock the large-scale unwieldy and often monotonous property developments and the macro-planning vision. This is where the focus of attention on the commonplace in architecture once again becomes a political act; one that maintains the legitimacy of individuation within the sea of generic commercial building styles. One symbolic image underscores this fascination with the informal as a liberating force: In the government-funded project initiated by the Stiftung Bauhaus in Dessau, Celula Urbana, the Bauhaus constructed a white modernist structure right at the heart of the favela Jacarezinho in Rio de Janeiro. The model will place the urban structures and the economic and socio-cultural demands of the favela ("city within a city") on a par with those of the "official" city. The brief aims at a model of intervention which embodies high standards of design and technology and generates and represents networks, communication and exchange with the "reputable" city.

The clean four-storey building rises out of slum as a testimony to a modernism re-aligned with dirty realism. This modernist legacy of functional style fused with the shack aesthetic is a pertinent symbol for current practice.

"Shamelessness can become useful" –
Yoshiharu Tsukamoto, Atelier Bow Wow

In the post-bubble economy of Japan where space is highly contested, the small-scale adaptations and programmatic hybrids are for the architects Yoshiharu Tsukamoto and Momoya Kaijima vital indications of individual freedoms within financially restricted and compromised situations. *Pet architecture* is a long-term documentation of micro-architecture and its creative typologies. Each minature intervention in the city's fabric is lovingly charted:

AOYAMA REAL ESTATE AGENT Real estate agent office in-between two buildings. It has only 0.8m in width but 10m in depth. It cranks in the middle following the shape of external stairs of next building. The minimum frontage to the street is carefully used as entrance and signboard on it.

OOKI BICYCLE SHOP Bicycle shop left over between sidewalk and ten-storey-high building. The depth measures only 0.6m but enough for setting up a computer and various parts of bicycle because shop owner works on the sidewalk. The practice of bicycle shop on such tiny site is also the protestation against the gentrification of the area.

These structures, according to Bow Wow, are cocking a snook at the gloss of large-scale capitalist development and resolutely holding fast to traditional Japanese forms – even upholding the scale and the intimacy of the traditional family house and the village within the metropolis. By preserving and intricately describing the small-scale spatial adaptations the architects are taking a stand in resistance to the relentless tabula rasa of commercial planning at Tokyo's core. Together with like-minded Japanese architects they are developing a laboratory for new architectural styles beyond conventional notions of beauty. This strategy of reading the city will then be applied to other contexts in Europe: "If we can't try to turn 'disgusting buildings' into resources, then there is no particular reason to stay in Tokyo. Surely we can start to think about how to take advantage of them. Rather than trying to run away. Shamelessness can become useful."

Such practice of extracting and developing new styles from Balkanisation, or from made-in-Tokyo hybrids and pet buildings, is currently drawing attention within the profession in the same way as did Venturi, Scott Brown and Izenour with *Learning from las Vegas* and the shed architecture of signs. William Gibson perhaps foretold the potential force of such strategies in his opening sentence of *Neuromancer*: "in an age of affordable beauty, his ugliness was legendary".

PRAGMATIC UTOPIAS – ITINERANT ARCHITECTURE

The present generation of itinerant architecture – short-term, adaptable, even viral, which resolutely resists the generic solution, has therefore established a foothold in quite a different economic reality. The emphasis upon self-sufficiency, low-cost materiality and that of carving out one's own human-scale utopian project is as the architectural critic Simon Sadler has noted "the most radical sub-species of building: it is restorative, stitching together spaces torn by insensitive projects, and planting seeds of occupation in land left as waste".

This hands-on tactical design, such as the buildings nailed together from recycled material that the architect/artist partnership Folke Köbberling and Martin Kaltwasser built up overnight at the fringes of urban housing projects takes its cue from the survival strategies of subsistence economies: the Turkish Geçekondu, if literally built within 24 hours, can claim ownership of a piece of land.

The architects N55 from Denmark, who have been building modular living structures for over 20 years, similarly seek out urban areas where there is less control – "small pockets in the cities that are in some kind of transition, rooftops, or old industrial areas". By suggesting ways of living that need little maintenance and tapping into a world of construction methods that have hitherto remained untouched, they combine design with a message of concentrating pragmatic, workable utopias: "Our work is about suggesting ways to exist within as small a concentration of power as possible."

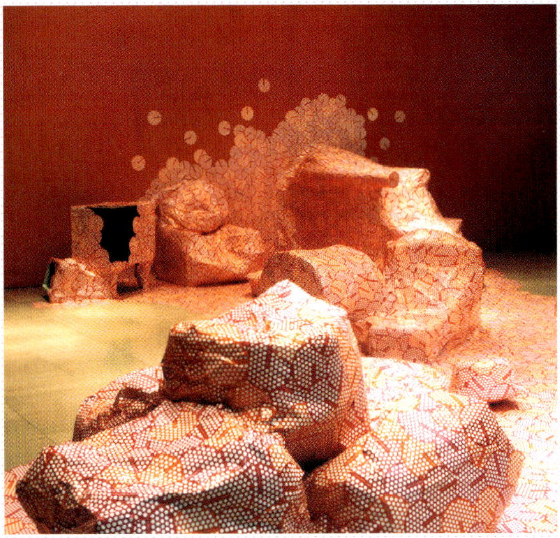

El Ultimo Grito, "Tagged Environment". London-based designers Robertoi Feo and Rosario Hurtado use 3D graffiti to re-conquer architectural spaces.

NICHE-ECONOMIES

Atelier le Balto call themselves garden architects, yet their temporary gardens in Berlin, planted in gaps and infill sites and making use of weeds that have grown out of all proportion (*Götterbäume*) exemplifies a niche-economy, landscape gardening as ingenious waste-management, with an entirely new aesthetic. By using materials such as slag and found rusted railway lines, they lend herbaceous borders a whole new meaning. This residual urban design practice has no illusions of grandeur and delights in the potentials of leftover spaces as a pre-requisite for reactivating a public awareness for the urban landscape at the margins of perception.

Such micro-topian actions have often evolved from a frustration with the apathy that increasingly sophisticated consumer-driven western economies have induced in the citizen-as-consumer. Why not turn the consumer to producer of his own environments? The generation raised on IKEA – according to the design partnership El Ultimo Grito, Rosario Hurtado and Roberto Feo, believe that by buying cheap they are beating the system that manipulates desires. And yet the resulting interior spaces are nonetheless immediately recognizable as part of the IKEA economy brand.

El Ultimo Grito therefore propagate *DIY as Design It Yourself* – exhorting the addition of another layer of responsibility to environments and objects as a safeguard to individuality, since, they maintain, "Acceptance of what's on offer means acceptance of someone else's values, definitions of normality, culture or desires." With their viral landscapes made of layer upon layer of stickers and tape they call "tagged environments" they are appropriating and rearranging fragments of the city to give them another life. Cedric Price's non-plan of 1969 is given its playful expression with minimal means through design propositions such as these: "We like to imagine cities that are not subjected to planning, where people do whatever they feel is necessary: where they create their spaces according to their own needs, reflecting their individual creative instincts." With stickers and tape, derelict objects are wrapped, "swarmed" and given another layer of meaning.

The transformation of existing spaces is about getting involved and replacing them: The Taiwanese architect Li H. Lu foretells his *interbreeding fields* as a provisional space generated by simple, standard-sized wooden beams. The viral architecture as an interbreeding and interfering laboratory re-establishes, as do Ultimo Grito, the necessity of play and the layering of space as generating dialogue, and enabling perceptual shifts. These are generated by the participant, as much as by the construction. Each manifestation of interbreeding fields, whether bridging a street in Taiwan, or adding play-and-view structures to a disused palazzo in Venice, becomes a reintroduction of a public, collectively enhanced perception.

THE INVENTORY OF THE STREETS

As if acknowledging the futility of actually building within fragmented edge scapes of contemporary cities, the micropolitics of design and architecture proposed and carried out by the *Talking Cities* protagonists instead define the need for a new politics of place. This is the architect and designer as social mediator.

Jürgen Patzak-Poor, working with Building Initiative within Belfast where, in the wake of the peace process, the generic commercial developments are indifferent to local contexts and the need for public spaces of discourse, declares that "contested meanings, uses and ownerships are in the nature of urban space and urbanisation in general. Architects can no longer propose solutions to these, as there can be no spatial solutions to social problems, but can learn to cope and live with them, as there can be social solutions to spatial problems".

Thus *Talking Cities* implements a collective desire to regenerate, recycle, re-engage the contested, the marginal, residual spaces of the contemporary cityscape and, in doing so, pays homage to the provocative questioning of the discipline itself by architects such as Cedric Price. "Sitting in our pods surveying the Ruhr, attending jamborees in the Rust Belt, we might address that old triumvirate of questions – who are we? Where have we come from? Where are we going? and to do this would, one hopes, return us collectively to the public, political realm." (Simon Sadler).

The act of gathering and exhibiting contemporary positions on architecture and urban design thus becomes a conceptual and practical model for a critical cultural aesthetic – a kind of anti-architecture, a site for multiplicity in architectural and design practice, and a form of improvisation.

Interbreeding Field, "Non-Side Zoom: Open Reality," Tainan City, 2005. With the assistance of his students, Taiwanese professor of architecture Li H. Lu creates new types of spaces in order to develop architectural concepts through practical experimentation.

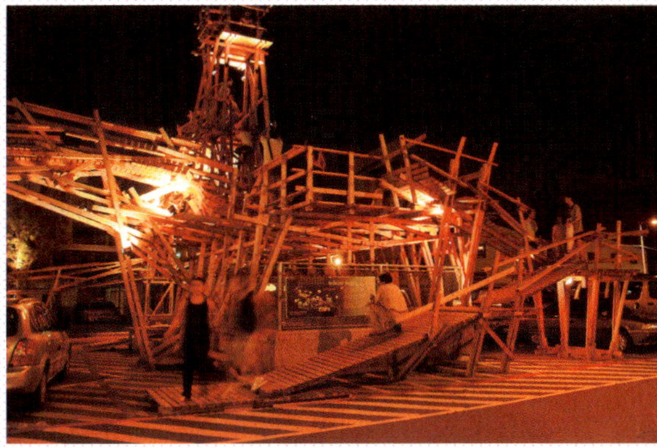

The London-based architects Takero Shimazaki and Yuli Toh best sum up the contemporary design positions that make *Talking Cities* a kind of urban microcosm: "The idea of a city as a vehicle for social interaction has metamorphosed from a physical conglomerate to a digital framework. The Fun Palace foretold this new talking city, where we now agree to meet. Where then, is the architecture of the physical city? We propose it has dispersed to the personalized place and architecture can now become unapologetically non-universal but intelligently specific to the individual – the Uncommon Solution".

Most of the herein quoted texts and procjects are widely documented in:
Talking Cities, The Micropolitics of Urban Space – Die Mikropolitik des urbanen Raumes, Francesca Ferguson (Ed.), Birkhäuser, Basel, 2006

PETER REED

GROUNDSWELL
CONSTRUCTING THE CONTEMPORARY LANDSCAPE

The new urban public landscapes that comprise this exhibition did not exist as public space half a generation ago. Nearly every significant new landscape designed in recent years occupies a site that has been reinvented and reclaimed from obsolescence or degradation as cities in the postindustrial era remake and redefine their outdoor spaces. *Groundswell: Constructing the Contemporary Landscape* portrays the surge of creativity surrounding the contemporary designed landscape by presenting a diverse selection of plazas, public parks, and urban sectors that have recently been completed or are in the process of being realized. The survey reveals the diverse ways that cities confront change by revaluing and reprogramming their spaces, ranging from a small urban square created in the aftermath of a terrorist bombing to the daunting task of transforming America's largest landfill – where as many as 29,000 tons of garbage were dumped daily – into a wildlife sanctuary and place for recreation.

The projects in the exhibition illustrate many kinds of sites, a multivalent term that includes the physical properties of the ground, its history, and a program for the new landscape. "Urban Stages" explores innovative reinterpretations of the traditional town square. The designers of these projects have endeavored to create unique visual and symbolic identities by various means, such as the selection of materials, trees, and plants with meaningful associations, the incorporation of iconographic elements, and the recovery of historical traces. Projects in "Simulations of Nature and New Topographies" explore the relationship between artifice and nature in various ways, including the sculpting of

In the postindustrial age, cities are endeavoring to redefine and redesign public space. It serves as a catalyst for urban development and quality of life, declares Peter Reed, thus in the end, our approach to it also reflects our culture. Reed's exhibition, entitled *Groundswell: Constructing the Contemporary Landscape*, is a portrait of creative potential in terms of the design of today's landscape. It presents a broad range of squares, public parks and city areas that are still in the planning stage or have just been completed. In his essay on the exhibition, Reed stresses the variety and originality of novel uses and gives some designers the opportunity to comment on their various projects and methods of approach. The "Urban Stages" body of topics addresses innovative reinterpretations of the traditional town square. "Simulation of Nature and New Topographies" investigates the relationship between

new topographies that are inspired by natural environments. Examples of such simulations range from a prairie garden constructed over an underground parking garage to a new coastal park inspired by dunes built on landfill. Among the most pervasive landscapes encountered in our postindustrial era are abused and polluted sites that present challenges ranging from the remediation of toxic environments to the defining of new programs for these inhospitable places. The compelling transformations of a number of such sites, which include a defunct steel mill, a former military airstrip, industrial waterfronts, and sanitary landfills, are explored in "The Bad and the Beautiful". Their former uses exhausted, architects, landscape architects, and urban planners are asked to envision their postindustrial transformation into places for leisure activities and redevelopment.

The reinvention of the sites surveyed here are a measure of artistic creativity and of our changing attitudes toward natural and man-made environments. The formal diversity in these projects reflects not only each individual designer's aesthetic and theoretical ideas about landscape – an art of horizontal surfaces and systems, impermanence and change – but also the way a design responds to a particular site. Inasmuch as landscapes are made and remade, the question of whether to reveal or conceal the traces of a site's history – be it natural, industrial, or political – defines its genius loci and ultimately plays a role in its form, meaning, and cultural significance.

Included here is a small selection of projects exhibited in *Groundswell*. These are accompanied by statements from several of the designers whose works are represented in the exhibition. Their statements are intended not so much as an explanation of a particular project, but rather to convey a broader range of ideas and concerns regarding the designers' approach to constructing contemporary landscapes. The matrix of ideas and

artificiality and nature. It talks about the formation of new topographies that are inspired by natural environments, such as a prairie garden that was constructed on top of an underground parking lot. "The Bad and the Beautiful" presents concepts for the reuse of closed industrial facilities. One example of an almost impossible project is the plan to transform America's largest waste disposal site, onto which over 29,000 tons of garbage are dumped daily, into a nature reserve and recreation area. In all concepts and projects, it is important for Peter Reed whether traces of the past are to be laid bare or covered over.

images reflects some of the ways that designers give shape to inert and living materials – rich in textures and colors and enhanced by light and water – to create meaningful spaces. Because of the role public space plays as a catalyst for urban development and in the quality of civic life, how these spaces are treated is ultimately a reflection of our culture. As such, the constructed landscapes in this exhibition are causes for optimism. --- *Peter Reed, Senior Deputy Director of Curatorial Affairs, The Museum of Modern Art, New York*

We need to grasp the truth that our landscape is a built artifact, and as such, must be designed, and therefore is an opportunity and an appropriate place for the expression of personal aesthetics and ideas. The visual quality of our environment defines us as individuals and citizens of a certain place. It becomes internalized and part of our self-image. It defines who we are individually and collectively and ultimately becomes what we are remembered by as a culture.

My definition of landscape is not limited to our natural environment, or to just parks, waterfronts, and plazas, but more importantly focuses on the landscapes we are building for ourselves in which we work, live, shop, commute, and generally exist. In spite of my reputation as a nature-hater, I am simply misunderstood. I am protesting this disjuncture between word and deed as it concerns nature and interferes with our ability to create positive environments for both man and beast. --- *Martha Schwartz, Martha Schwartz, Inc., Cambridge, Massachusetts*

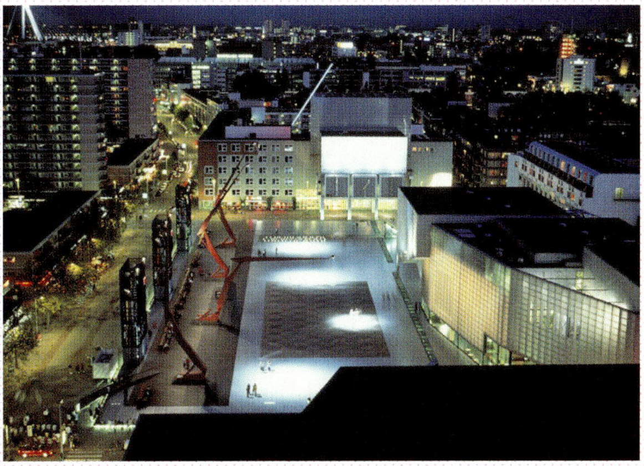

Adriaan Geuze, West 8 urban design & landscape architecture, Schouwburgplein (theater square), Rotterdam, view by night

Modern-day citizens can use landscape and nature as one way of defining their place in the world. They identify with the landscape, which is an essential basis for urban design. The elements of landscape become the source material from which the structure and soul of the city is derived.

The landscape architect should protect cities from the horrors of junk space and ersatz nature by utilizing simple but clear landscape elements. In addition, he needs to create enclosed spaces; condensed, spiritual oases that allow escape from everyday life by offering the sublime illusion of a fantastic other world. --- *Adriaan Geuze, West 8 urban design & landscape architecture bv, Rotterdam, the Netherlands*

In a history of rapid transformations that typifies our modern mega cities, landscape has become an ambivalent field of experimentation, ranging from the exquisite garden sanctuary to the despicable peripheral wasteland. There is, therefore, no longer a single approach to landscape design, but rather a multifaceted response to ambient conditions and history. My job as landscape architect is to assess to what extent I may add a new layer to the strata already in place. A landscape vision requires patience. The inherent meaning sown in each project, still dormant, probably needs another couple of decades to reveal its full potential. --- *Christophe Girot*

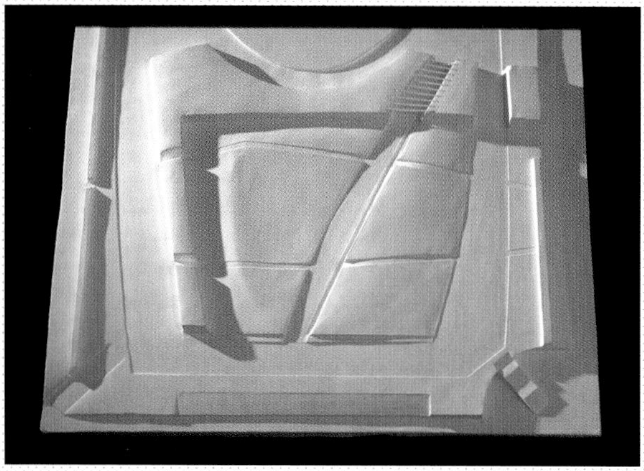

Kathryn Gustafson, Gustafson Guthrie Nichol Seattle, Washington with Piet Ouldolf, Hummelo, Netherlands and Robert Israel, Robert Israel Inc., Los Angeles, California Lurie Garden, Millenium Park, model of the concept, bird's eye view

Open space is as important to civic, cultural, and modern social life as interior space. The designed landscape can be as capable of commemorative expression or mystery as any façade or other architectural form or dimension. It is the public open space formed for function only, filled with purposeful but artistically bereft roads, parking, and service spaces, for instance, that carries the message of indifferent ugliness, thereby tarnishing the hopes of modernism to the degree that modernism is felt to have in fact failed. A large part of that failure lies in site planning and open space areas, the public realm of cities and towns. --- *Peter Walker, Peter Walker and Partners, Berkeley, California*

What is the environment for the people who live in a postindustrial society? Landscape design has begun to join with technology and social science as well as art. But it is also true that the search for beauty that simultaneously looks to intuition and expression exists here. We must lay our landscapes of self-realization within the social scene that seeks justice for the public, and establish a design that shares these. This is where we stand today. --- *Yoji Sasaki, Ohtori Consultants Environmental Design Institute, Osaka, Japan*

The opposition between the rational and the organic that structures the history of landscape design has characterized the history of several disciplines, from philosophy to urbanism... It is in the overcoming of this opposition that we think the possibility of an emerging landscape – and city and architecture – may exist. We do not need to resort to the "pure" artificial forms, to detach geometry from the contingencies of the topography, wind, or vegetation; we can adjust them through computer modeling of the forces. Only by applying these techniques rigorously will we be able to meet the challenge of creating a new discipline across the natural and the artificial, the rational and the organic. --- *Farshid Moussavi and Alejandro Zaera-Polo Foreign Office Architects, London, England*

›
Detail of Water Garden, Bordeaux Botanical Garden. La Bastide, Bordeaux, France, 2000–02. Catherine Mosbach, Mosbach Paysagistes, Paris, France

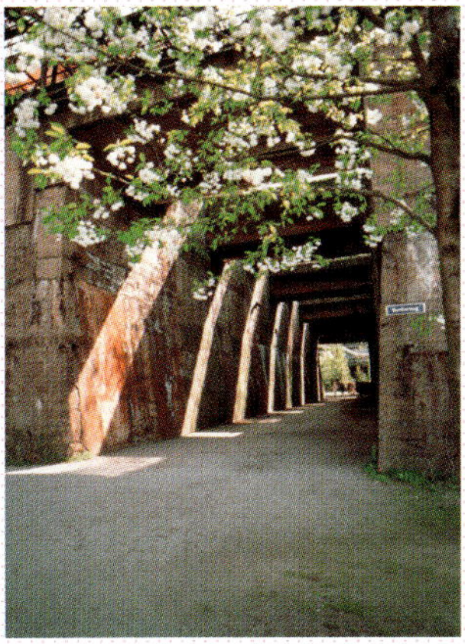

Peter Latz, Latz + Partner, Kranzberg, Landschaftspark Duisburg-Nord, Internationale Bauaustellung (IBA), Emscher Park, Duisburg 1990–2002, corridor between two former ore bunkers

Before I begin designing, I try to physically feel the essence of a site. A site has a soul – a physical past, a cultural past. I am interested in the manipulation of space and the way that terrain, which has mass, can focus our engagement with the landscape as a figural body that is strong, spare, and sensual. The connection to a site is discovered with thoughtful movement and is guided by volume, textures, and illumination. The connection is further refined through the definition of what is solid and what is fluid, through exploration of the culture and constraints of place. --- *Kathryn Gustafson, Gustafson Guthrie Nichol Ltd, Seattle, Washington*

In my projects I am always interested in the "genius of the place" rather than in the genius of my office. Even motorways or industrial wastelands can be filled with a new spirit and can be made worth living by keeping visible the spirit of the existing site. Landscape and open space contain a wealth of information layers. These layers of information can exist physically, they can be visible or invisible, they can be abstract like cartographic grids, or remembrances – lying within or out of the place. The challenge is to make the right selection, to liberate our senses and to be open to new impressions. --- *Peter Latz, Latz + Partner, Kranzberg, Germany*

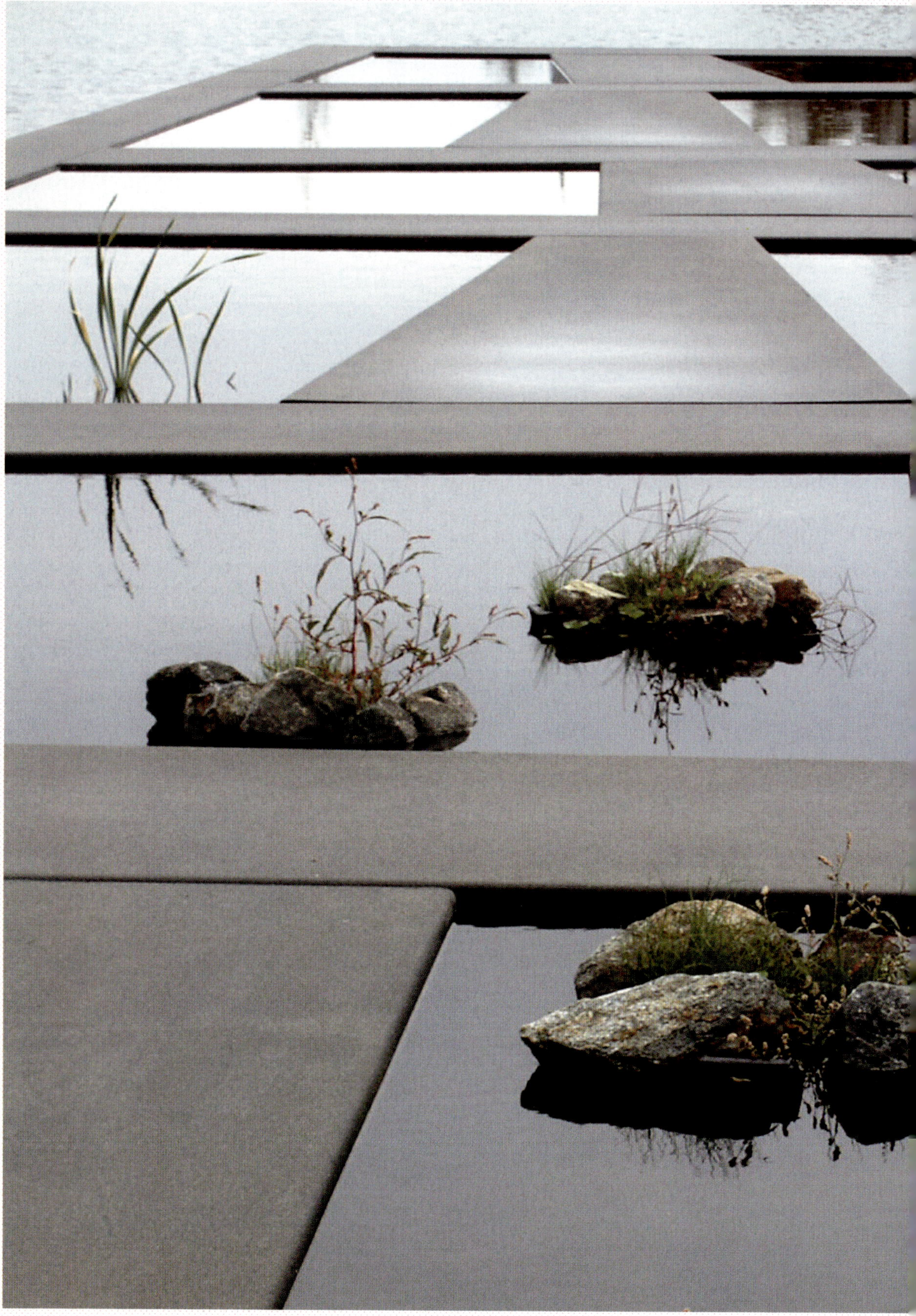

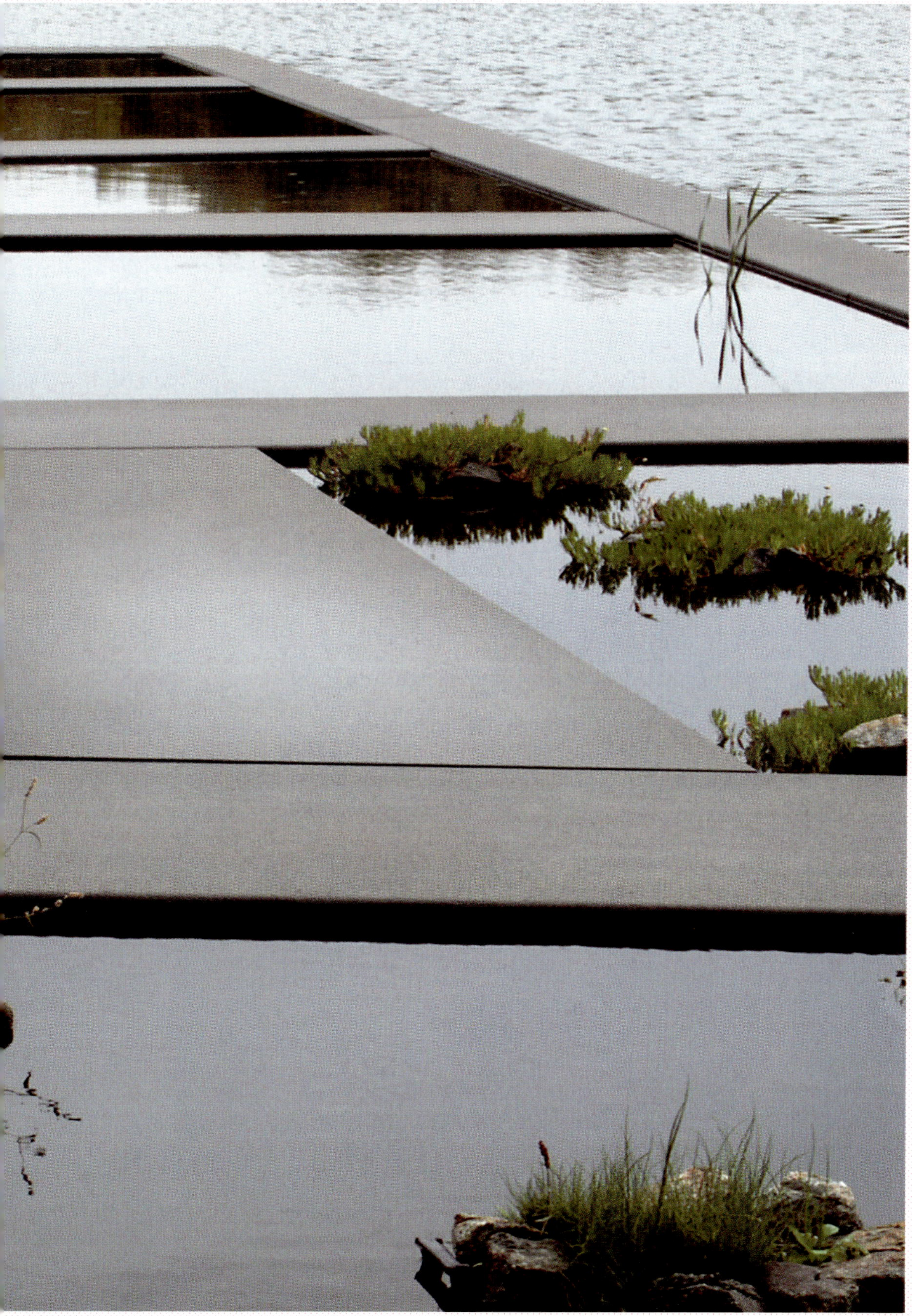

Hargreaves Associates, San Francisco, California and Cambridge, Massachusetts, Crissy Field, Presidio National Park, San Francisco, California 1994–2001, prior to reconstruction the grounds were used as a car park (on the l.) and then as an in part flooded wetland zone (on the r.).

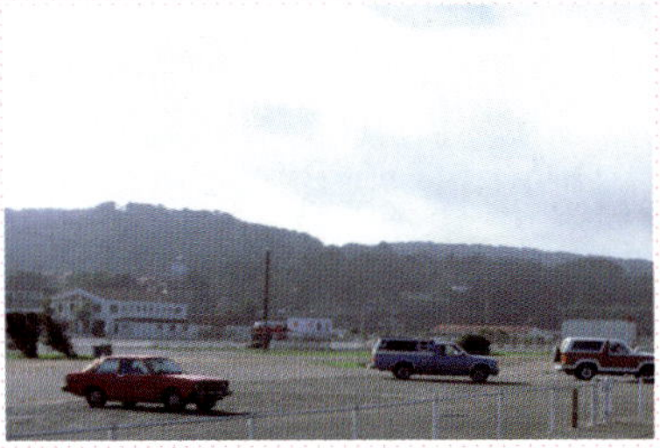

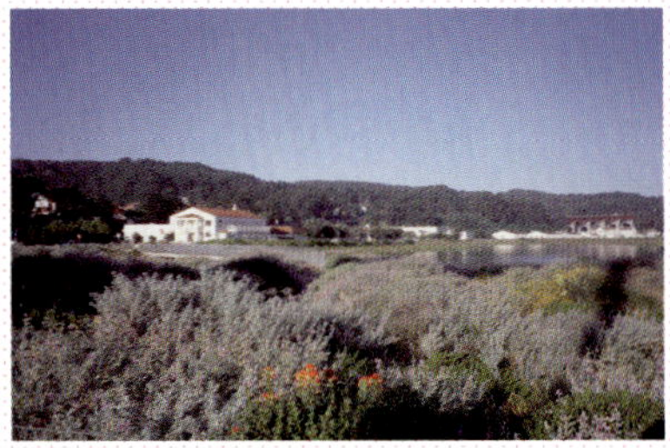

We design to clarify the physical and cultural identity of a site, opportunistically integrating the disciplines of landscape, architecture, and infrastructure engineering. The integration of these disciplines has been central to how we design. We do not believe in the customary distinction between art and nature, design and ecology, but believe that art and nature are bound together in reciprocity. Selective removal, tactical excavation, and strategic additions transform the perception and experience of architecture and the site. Our goal is to avoid the pretense of imagining architecture only on a neutral site or the confused agenda of imitating natural forms. --- *Marion Weiss and Michael Manfredi, Weiss/Manfredi Architects, New York, New York*

Showing landscapes in the heart of the city meant presenting a world, a sphere of nature, that could speak to the visitor and take him out of himself. For me, the "culture-nature" relationship is rooted in childhood when I turned away more or less consciously from all that was intellectual or linked to education. Since then, in my professional activity, it feels like I am doing exactly the opposite. Meetings with like-minded people have been essential to me in my intellectual development, which has always been guided by my emotional life. Between emotion and knowledge, the Bordeaux garden seems inexhaustible to me. --- *Catherine Mosbach, Mosbach Paysagistes, Paris, France*

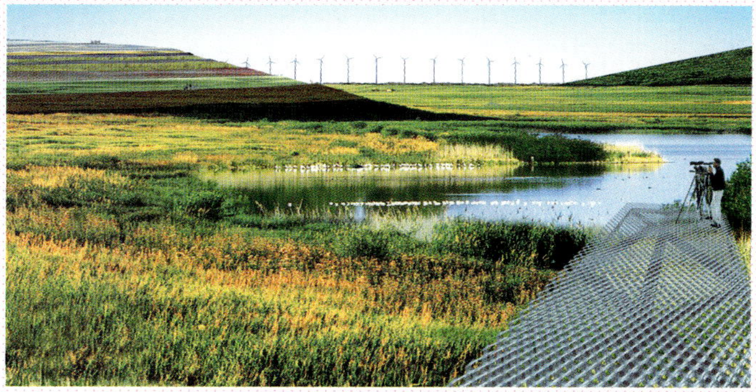

James Corner, Field Operations, New York, New York. Fresh Kills "lifescape", Staten Island, New York, 2001–05; building work is planned to start in 2007, rendering

Site is the fundamental building block of any landscape. Typically abandoned and/or contaminated, however, today's sites often require vast transformations in order to be returned to the public realm. At issue in these transformations, then, are the conscious qualities of site character, human activity, sustainability, and site-making. In developing site character we are asked to engage in notions of place, form and pattern, spatial volume and surface character. And through site design we must bring to life the delicate balance of efficiency and theoretic ideal that is sustainability and natural process. Emergent from all these considerations is the ultimate concern: to engender in the public realm those great places that capture the minds and hearts of humanity, and over time reflect the soul of our civilization. --- *George Hargreaves, Hargreaves Associates, San Francisco, California, Cambridge, Massachusetts, and New York, New York*

Contemporary urban projects demand a new kind of synthetic imagination – a new form of practice in which landscape, architecture, planning, ecology, engineering, social policy, and political process are understood and coordinated as an interrelated field. The synthesis of this range of knowledge bases and its material embodiment in public space lie at the heart of strategic landscape practice. Creative strategy, process, temporality and strong physical design are crucial for generating new natures, new publics, new programs, new forms, new urbanisms. --- *James Corner, Field Operations, New York, New York*

The exhibition *Groundswell: Constructing the Contemporary Landscape* was organized by The Museum of Modern Art, New York

THE OBJECTIVE FORM OF THE ZOLLVEREIN COAL MINE WAS CONCEIVED NOT FOR ETERNITY BUT FOR IMPACT. IT WAS TO GENERATE PRESTIGE. WITH THE CLOSURE OF THE MINE IN 1986, THE ORIGINAL PURPOSE WAS LOST. ITS NEW USAGE WAS BASED ON THE CONSERVING AND RESTORATIVE CONCEPT OF PRESERVING A HISTORIC SITE. IN THE MEANTIME, FURTHER ARCHITECTURAL HIGHLIGHTS AND INTELLIGENT NEW USES HAVE BEEN ADDED. FOR THE SPRAWLING URBAN LANDSCAPE OF THE RUHR DISTRICT, THE OLD MINE WITH ITS LANDMARK ARCHITECTURE SERVES AS A PROJECTION SURFACE FOR REINVENTING ITSELF AS A CULTURAL REGION.

THE ZOLLVEREIN COAL MINE

--- A PLACE OF RENEWAL ---

FABIAN WURM

A SYMBOL OF STRUCTURAL CHANGE
NOTES ON THE ARCHITECTURE OF SHAFT XII

The mining magnates were as proud of their pits as the Pharaohs were of their pyramids. Architect Herbert Gunia mentions the Zollverein, part of UNESCO's World Cultural Heritage, in the same breath as the seven wonders of the world. The man waxes euphoric when he starts talking about the coal mine, the building owners and the building masters. "Schupp was unique, compared to him we are just small-time craftsmen." The two Zollverein planners, Fritz Schupp und Martin Kremmer, have remained role models. Their work encompasses more than 40 mines as well as churches and coking plants, continuous casting and switching plants as well as chemical and cement factories: a mighty heritage.

Today, Gunia manages the Essen office of Franz Schupp, who died in 1974. "The company is now in the third generation," as he quickly adds. He considers himself a descendant of the founding father. For many decades he has guarded the plans of the legendary industrial architect born in 1896 in Uerdingen and organized exhibitions throughout Europe with the photographs from the estate. The name of the company founder, however, had to disappear from the company logo. The master himself wanted it that way. "The lads only mess things up anyway," founder Schupp is supposed to have muttered. He wanted to distance himself from it all, but the heirs wanted to be grandchildren. In the office's profile we read: "We have consistently advanced the office in keeping with the tradition Schupp established."

Gunia and his partner are specialists in industrial architecture. Just like Fritz Schupp and his co-pilot Martin Kremmer (1895–1945) before them, they have also designed

Design and art, dance and theater between cooling towers and coal washing facilities, boilers and compressors – Zollverein is a place of contrasts, a symbol of the renovation of an entire region. Zollverein colliery's 55-meter-high double headgear always crops up in conversations about structural change in the Ruhr District. It is proudly called the "Eiffel Tower of the Ruhr District", the "New Age Colossus of Rhodes", while the whole shaft plant is even celebrated as a "synthesis of the arts in the Bauhaus style". Even though these names may be misleading, one thing is clear: Zollverein amazes us with its sheer size and power. It was once the most efficient and most modern of all coal mines, now it is the largest industrial monument in the international arena – nothing but the best. When the steel cables of the winding tower were tightened on February 1, 1932, the press swiftly announced the first record:

mines, coking plants, and blast furnaces. Structural change in this sector has not passed this old established office by without leaving a trace. Pits have been flooded and the fire in the coke furnaces blown out. The architects have even witnessed the Dortmund Kaiserstuhl coking plant being dismantled: Chinese workers took apart one of the largest plants built to the Essen architects' design in order to ship it back to China and re-assemble it there. That was in 2003 and 2004.

The great era of industrial building in this sector has been and gone. What remains are reminiscences, memories of people and places. In well-off southern Essen, Fritz Schupp dreamt up plans for the giant industrial plant in the structurally weak north of the industrial area. He lived and worked in Rüttenscheid, Stadtwald and in Bredeney when he was not busy in Berlin with his office partner Kremmer. He had a view of the Krupp family's park around Villa Hügel from the building where he lived and worked – Ruhrstein 15, still a prime address today. The best area in the Ruhr District largest city. Some of the developers lived in the vicinity. And Schupp's descendants also run their office in Bredeney.

The no. 107 streetcar links the north and south, the villa district and the mine estates. Essen has two sisters – the South, green, and the North, gray. Even though much has changed, the cliché is not entirely false. On the one side, villas and forests which extend down as far as the Ruhr, on the other, workers' houses and industrial wasteland all the way to the Emscher river. The Ruhr Valley is a recreational area, while the Emscher banks (for all those dreams of a "blue Emscher") is the area's back yard. It is almost as if we were witnessing two different worlds. Linked by Line 107, which travels on a narrow track from Bredeneyer Kreuz past the Zollverein to the last stop, Gelsenkirchen's Music Theater. A trip through space and time, a "special kind of city tour" – at least that is how the rapid transit operator at Essen Verkehrs AG has labeled it. In August 2005, the "Culture Line" train was inaugurated. The specially compiled Cultural Route lists about 60 sights

12,000 tons of usable hard coal were extracted per day, that was four times the daily extraction of an average mine in the Ruhr District. Rationalization steamed ahead. The seemingly fully automated central shaft plant XII appeared to be enough in itself. There was nothing for foreigners in the complex. Even miners were not tolerated in the Ehrenhof, the impressive entrance area. The whole complex in Essen's Katernberg: a "forbidden city". A shaft staffed by ghosts, exemplary colliery, automated machinery to a large extent – what was Zollverein? Bauhaus architecture or industrial Baroque? A "miserable monstrosity", as author Erik Reger described the facility in 1932: "It was the German disaster, it was the intoxication with the gigantic." Zollverein architects Fritz Schupp and Martin Kremmer adopted an objective pathos. In its overall effect, the colliery complex is powerful, but in its individual parts, it is meager.

along a 17-km stretch, including two opera houses, the house of a philharmonic orchestra, even a voodoo museum, and the world cultural heritage, the industrial cultural landscape of the Zollverein in Essen's Katernberg sector. The transport company recommends the day ticket for EUR 10.20, which allows up to five people to travel on the line for the whole day. The high point is undisputedly the Zollverein coal mine.

The once soot-blackened settlements are brightly painted. Pithead XII's bi-partite 55-meter-high headgear raises its head between fields and mining colonies. "The Ruhr Valley's Eiffel Tower", a "modern Colossus of Rhodes" as it is called, or sometimes the whole complex is referred to as the "Cathedral of Work" or as an "entire artwork in Bauhaus style". One description is as confusing as the next. Once again it becomes clear that whenever it is mentioned people endeavor to use a superlative: the largest, the most efficient, the most modern and the most beautiful of all coal mines. The Zollverein gives us something to marvel.

Chroniclers back in the days when it was built were not able to resist the fascination. "The winding tower stands, legs astride, over the modern mining structures," noted painter and typographer Max Burcharzt in his book, *Coal in the Ruhr*, in 1932. "Anchored in heavy concrete blocks, the iron struts stretch upwards supporting buzzing pulleys at lofty heights." Burchartz, in those days a Folkwang School professor, took the entrance between the two gatekeeper houses coming from the streetcar stop. "I approach the open gate, pleasantly moved by the tidy lawn of the mine's square which is more akin to a park." Sooty equipment and coal-laden wagons are nowhere to be seen. Today's visitors also look upon a neatly mown lawn area, a large courtyard bordered by strict geometrically shaped and cleverly staggered, monumentally-sized buildings. Faced with this surprising reception who would still think of coal, sweat and work mates?

Schupp and Kremmer raised a monumental order – and flexibility – as the principle of the day. Thus the individual façade elements of their buildings – the bricks in-between the steel frame sections – could be quickly removed when extensions to the facility were required. Monumentality and objectivity: For the architects, this was no contradiction in terms.

Other associations are desired. The mine square as a park, and here, again, admirers and critics have already made the comparison: The 20-building ensemble is not dissimilar to a feudal complex. Zollverein – a Versailles of technical progress? The square between gatehouses, transformer and switching house, mining hall and workshop buildings – a *Cour d'honneur*? Then take the boiler house in the square, which would correspond to the *Corps des logis* in a palace complex. The term "honorary courtyard" has at least, prevailed – much against the will of Schupp and Kremmer. They certainly did not wish their broad entranceway to be seen as a rehash of ostentatious Baroque forms once common in mining architecture. Less the aesthetics but rather the plant organization required individual sections to be arranged linearly, the symmetric solid-walled winding tower demanded an adequate division of space and building mass.

Not only the entrance area of the Zollverein's Pithead XII is organized on an axial pattern, but also the path to the boiler house which lengthens the right side of the courtyard. At the beginning it is bordered by the cubic buildings of the central workshop and the electrics workshop, further on by the buildings housing the high pressure and low-pressure compressors. Everything is conceived for maximum effect. The architects designed the visual axis with a central perspective, the technical facilities at its ends completing the view: The frontal visual axis is directed towards the mine with hall and two-legged winding frame, the diagonal axis to the staggered boiling house – once upon a time, the chimney rose up behind it. Originally, the architects wanted the cooling towers left and right of the chimney to create an even more impressive effect. Even the lights along the path are tapered to increase the perspective effect.

Bauhaus architecture or industrial Baroque? Already at the time it was built the Zollverein planners were said to have a penchant for representation and pathos. Fritz Schupp does not deny it. "If an architect today dared arrange a large complex in the form

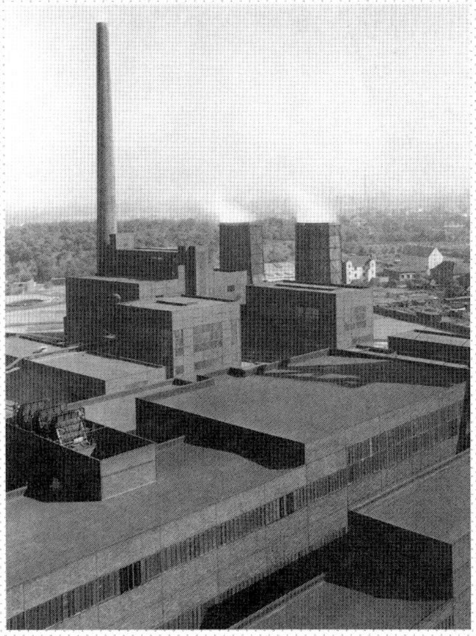

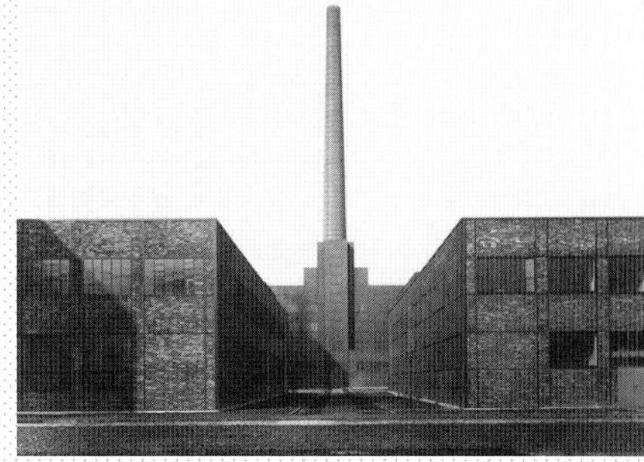

Commissioned by the architects: Albert Renger-Patzsch photographed Shaft XII in the 1930s. View from the coal washing unit across the hall with the picking conveyors to the boiler house (left). The boiler house with the chimney stack (center) that has since been torn down. The base of the headgear and coal bunker and coal washing unit (right).

of side wings and a central building around a courtyard he would probably be accused of being stuck in historical ideas. He would not allow himself to be misled by shallow criticism when the building best fulfilled its purpose with this kind of arrangement." The way the buildings are laid out is based primarily on the logic of technical processes, or at least that is how the architects saw it. The machinery used in washing coal limited the scope they had, in any case. What is evident here is that alone the purpose "ruthlessly perceived and correspondingly implemented" shall determine the design of the industrial plant. The architects and of course the developers also defined the purpose of the mighty entrance area clearly: It served the purpose of representation.

Mine directors and inspectors were indeed proud of the complex. They drove up in front of it and presented their astounded guests the more or less deserted, largely automated machinery. The workers came from behind, from the chomping of the adjacent primary shaft, Zollverein 1/2/8, over a narrow connecting bridge. In a way they entered the complex through the staff entrance, disappeared into mine bunkers or to the conveyor belt, the coal wash plant and machine houses. "The courtyard was the inner sanctum," recalls Hermann Neuhaus, who was a miner at Zollverein for four decades and today on behalf of "Zollverein Touristik" guides visitors through the mine settlements in Katernberg, "not a place for normal mortals".

Zollverein XII was conceived purely as a conveyor shaft. The miners entered the mine through the neighboring shafts of the Zollverein Group. Only a few workers were able to fit into the rooms above ground, machine operators, and apprentices who grabbed the dull stones from among the coal. And sometimes even they got in the way. As old Zollvereiners still like to report today, when Mine Director Friedrich Wilhelm Schulze Buxloh visited his modern exemplary plant the staff were asked to keep themselves out of sight. The spectacle of the loaded wagons which appeared to run automatically into the

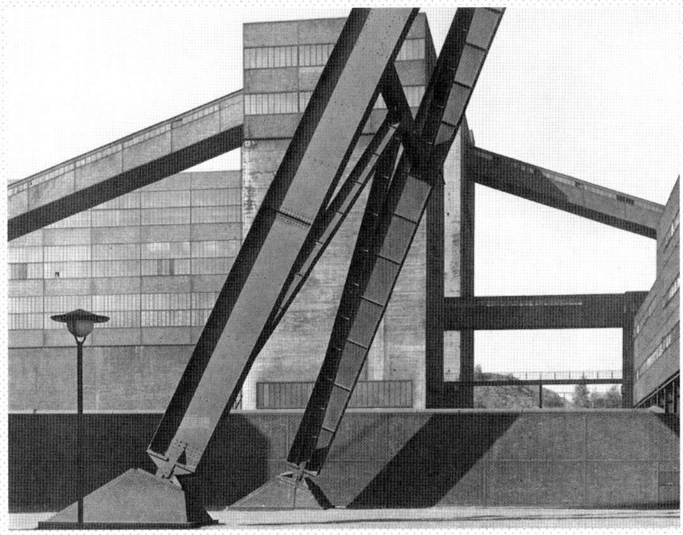

tipper, upside down to the other side one floor below, and return empty to the hall seemed to elicit respect and astonishment at the opening. The dignitaries took off their top hats, as another legend has it, when the *deus ex machina* in the form of a wagon rolled into the tipper. Technocratic mystification. People were speechless in the face of the merciless yet tremendously rational system.

Ghost mine, model pit, largely automated plant, the most effective in Europe – but what was the Zollverein XII really? This plant which was the last of the Essen mines to be put out of action in 1986, is a good example of the rationalization efforts of German industry, that much appears to be certain. Dreamt up in a short political and economic period of stability which was already over in 1932 when the mine took up operation. In 1926, Thyssen, Rheinstahl, Rheinelbe and Phönix, to which the Haniel Gründung belonged, joined forces to become the Vereinigte Stahlwerke AG, giving birth to the second largest steel group in the world. Its primary aim: To increase efficiency in order to hold its own on the competitive international steel market. This merger of the four West German iron and steel groups including their mining companies and processing operations received mixed feedback. Some critics considered Vereinigte Stahlwerke AG as the epitome of German monopoly capitalism, others as a failed attempt to revamp the economy of the German Reich. What is certain is that it was an example of a multi-divisional trust modeled on the US Steel Cooperation. Not just economists and engineers looked to the United States. Given the shortage of money in the German economy, investment needs could only be covered with the help of fresh money from America.

Only the Vereinigte Stahlwerke AG managed to acquire the necessary financial means to build a mighty plant like the Zollverein XII. Here we should mention the developers: first and foremost, Mining Director Friedrich Wilhelm Schulze Buxloh (1877–1959), the driving force behind the project in Katernberg; on the foundation of Vereinigte

›
Forbidden City: A photo by Anton Meinholz dating from 1934 long influenced the public perception of the complex. The Essen-based photographer showed the mine as an inapproachable monument. Access to the courtyard is blocked and the eye directed to the 55-meter-high double headgear. Meinholz emphasizes the symmetry of the entrance, the lawns, the winding gear, and the shaft hall, which still bears the original signage: "Zollverein". Four years later, the sans serifs had been replaced by fraktur letters spelling "Zollverein Schacht Albert Vögler".

Stahlwerke AG he became a member of its Board and was entrusted with managing the Gelsenkirchen Mining Group. Architect Schupp was his man. First contracts were awarded to the newly graduated architect back in 1919 – family ties are reputed to have played a role, although this cannot be precisely reconstructed. The mine director and the architectural partnership of Schupp und Kremmer worked together for many decades. That was the brief set by Albert Vögler (1877–1945), the Chairman of the Board who also had a seat in the Reichstag – initially championing Stresemann, and then Hitler – and during the War acted as Speer's chief representative in coordinating the production of military equipment in the Ruhr Valley. The mine management had erected a monument to him while he was still alive. As of 1938 the Albert Vögler Mine was prominently written in large Gothic letters on the mining hall beneath the headgear. Not until 1968 was the motto removed.

On February 1, 1932 when the steel cables were ready to roll, the press expressed surprise at how efficient the industrial complex was: 12,000 tons of utilizable coal per day, this meant four times the daily average of which other mines in the area were capable. It took only 83 seconds for the four skips to be loaded and unloaded and pulled from a depth of 620 meters – from the 12th seam to the suspended ledge above ground. The coal washing plant was able to process 800 tons of pit coal every hour. Another statistic which is surely less likely to be mentioned is that 500 miners were laid off at the Zollverein when the seemingly fully automatic central mine started operation exclusively for the purpose of coal extraction.

Technical capacity should correspond with rational realistic architecture. Bulky in its overall appearance, but matter-of-fact and sparse in the individual forms: The architects aspired to order, clarity and discipline as their benchmark – not to mention flexibility.

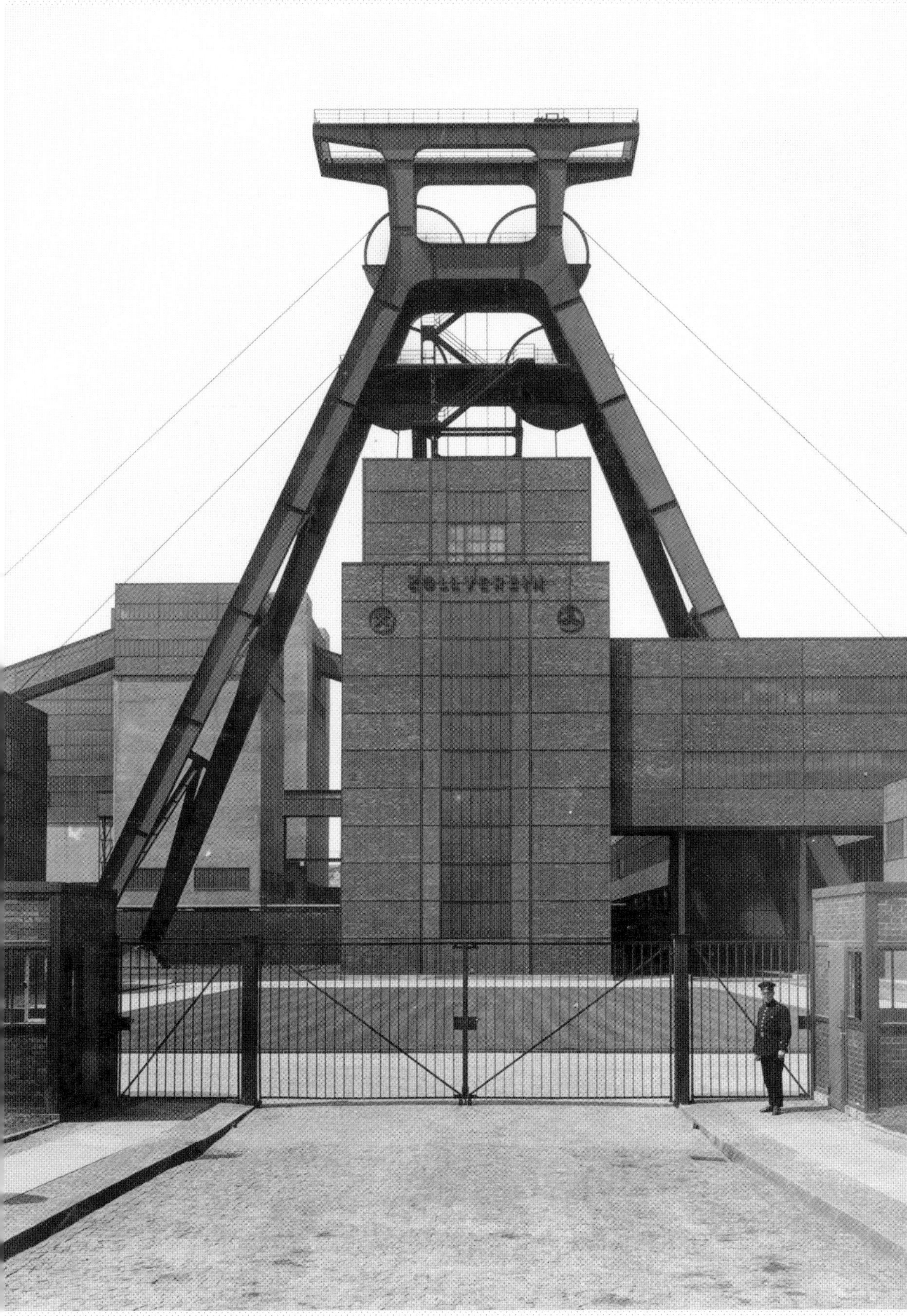

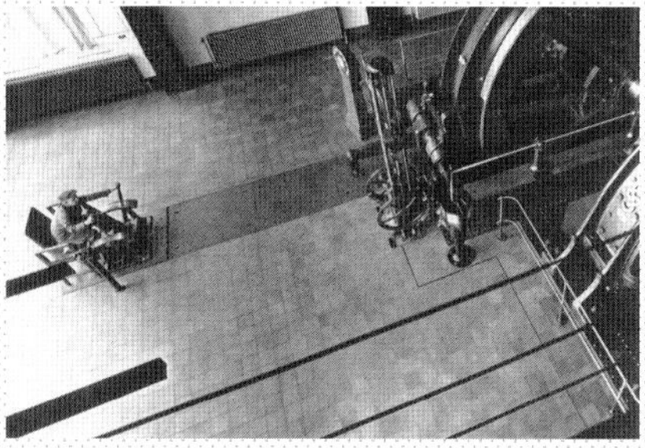

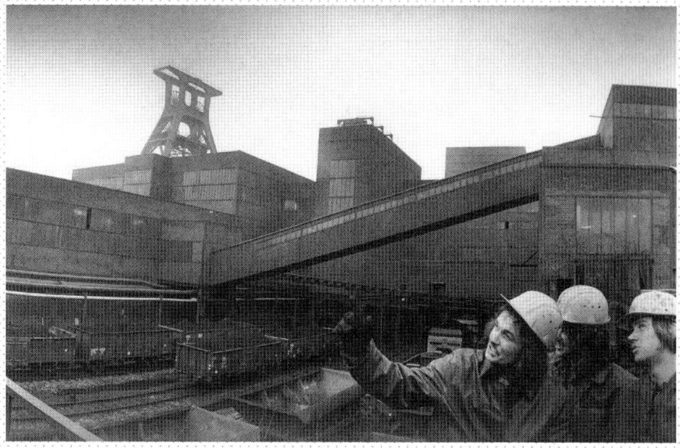

Pride of the colliery: Painter and typographer Max Burchartz created one of the few indoor shots from the time the structures were built. In his book "Kohle an der Ruhr. Eine Bildfolge mit erzählendem Text", which he edited in 1932 together with Walter Witzel, he presented the conveyor machine operators in Shaft XII (above). Barbara Klemm visited the plant twelve years before it closed down. Her photo came out in 1974 in the "Frankfurter Allgemeine Zeitung" (right).

Rationalization also meant for them that individual elements in their buildings had to be quickly replaceable when alterations in the technical processes or building expansions demanded. For this purpose they devised a flexible system: The steel-lattice form with double girders and U-shaped iron. The individual sections are framed by these thin steel profiles and can be closed with horizontal wire-reinforced windows flush with the wall or filled in with brickwork. The architectural details are all reduced to a few elements. The result is a well-proportioned, harmonious façade grid which extends across almost all the buildings. The buildings themselves appear as simple cuboids. A special artistic measure was needed to achieve this cubist effect: A parapet conceals the sloped roofs and the architects had the rain pipes laid behind the outer wall.

Although Schupp and Kremmer are not considered the inventors of steel-lattice frames (they characterized the image of factories and mines already at the turn of the century) they used this style in a new way. Their post-&-rail frames are merely a protective covering, as they have no supporting function. This covering, which measures exactly the width of a single brick (precisely 12 cm) – is "placed like a case over the contents, the operating equipment", as Schupp described it. The primary structure – heavy steel binders which carry the weight of floors, ceilings and when necessary machines – is to be found behind the outer covering, invisible from the outside. Should an industrial building conceal its construction and equipment? This sparked off some hefty debates at the time. Permeation of inside and outside. Transparency and constructive veracity were the buzzwords. Only visible structures could be considered honest works, the "denial of the wall" was the prerequisite for the "union of interior space with everything outside", as Walter Gropius expressed it. The barriers should be down, the machinery visible.

"Of course the machinery has become the starting point for the new perception of beauty," was Schupp's response to potential critics, but the relevance of a building was

New Objectivity images of the world of production: fourteen years after the plant closed down, the film "Metropolis" (director: Fritz Lang, 1927) was projected onto the façade of the shaft house. "Bizarre mechanical cogs started work on-screen; strongly rhythmical gestures commence..., glissandi the sensation as the cages with the slave laborers rushed downward into the oppressive underworld," is how the Recklinghäuser Zeitung commented on the open-air spectacle at the Zollverein.

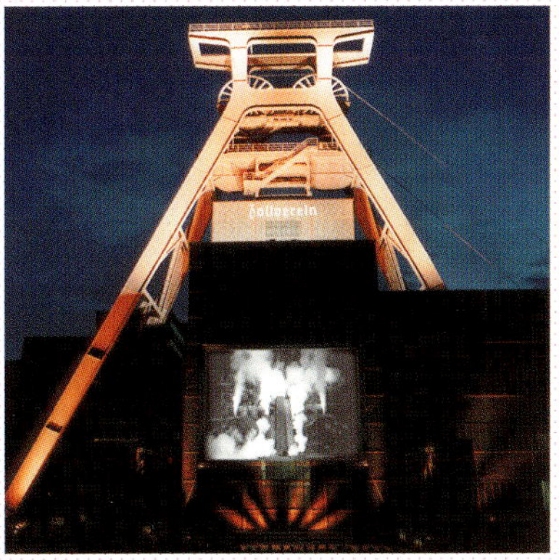

not dependent on it looking like a machine. Much to everyone's surprise the master industrial builder referred to the philosopher Immanuel Kant: Beauty is the form of the purposiveness of an object, "provided that it is perceived without a notion of the purpose." Schupp stated more precisely that the buildings should express the rhythm of the function which takes place within. This is what he wrote in what is probably his most important sentence published in a book with the syntactically unusual title of *Architect against/or/and Engineer*. That was in 1929. The young industrial architects formulated their program. The time of the great architectural manifestos was already over. Farewell avant-garde, New Objectivity has arrived. Praise of the engineers such as Le Corbusier still expressed at the beginning of the 1920s, have disappeared: "The engineer, advised by the rules of austerity and guided by calculations, shifts us into harmony with the laws of the universe (...) Engineers are trampling the dying architecture beneath them with their calculations." Engineer romanticism. No, the man of the machine Le Corbusier liked to wax mythical about does not appear as a Messiah to the Essen architects.

They have devised carefully calculated considerations, so far a bit wooden. On the basis of their own buildings for the coal industry, Schupp attempted to fathom the architect's scope in designing industrial plants whose function was fixed by the technical process. What role did the production engineer play when a new mine was being built? And what role did the architect play? This was important to clarify.

They by no means wanted to mix the tasks of the architect and the engineer. However, each needed to understand the work of the other. Yet it would have been "fundamentally wrong to expect that the architect be so trained in things of an operative nature that he could technically master the procedures in all their details," claimed Schupp. Also the architect was not there "to dress up a building". He was not just there to lend form but to give order to all the functional processes defined by the engineer, according to artistic criteria. To create "a higher order for the situation", according to Schupp. The architect is in a manner of speaking a conductor, one could interpret his role in this way. The hierarchies were also clear: The architect is the direct partner of the industrial tycoons. Significantly, the engineers of the Schupp-Kremmer buildings remain nameless.

In part, the work can be read like a protracted justification of the design of the Zollverein. And this is not surprising, because when the architects were conceiving their program they had already worked almost two years on the design and realization of Shaft XII. Here at the Zollverein the combined effect of architects and engineers was supposed to take shape on a grand scale. All their industrial buildings up until this point, clearly influenced by Expressionism, were hardly more than preliminary exercises.

Zollverein put the rule to the test: At long last architects were to be able to have an influence on the layout of technical plant during the planning phase for building a mine complex. The participation of architects in the building of mines and coking plants was not taken for granted in those days.

The architects painstakingly sought to defend their design for Europe's largest shaft. They wanted no false associations to arise. The boiler house a cathedral? "The architect will not have some heaven-aspiring cathedral in mind when working on his design," wrote Schupp and explained that industrial plants now required a certain size. Objectivity was the rallying cry. "Other groupings of boiler and compressor houses are conceivable but none of them would display the tight clarity of ground layout and be as economical in their use of space as when organized around a courtyard."

Monumentality and objectivity – that was no contradiction for the architects. True, their plants have little in common with Bauhaus architecture. "Today's buildings have turned their back on the representative symmetrical façade," decided Walther Gropius in 1925 and conceived his Bauhaus buildings as consciously asymmetrical constructions. According to the Bauhaus founder, "One has to go around the building to grasp its physical nature." With the Zollverein, however, this is hardly necessary. The viewer can quickly form an image from central positions: The spatial disposition of the complex and size of the building is easy to recognize. Social factors too, such as those taken into account by Bauhaus founder Gropius in his factory designs, appeare to have hardly interested Schupp and Kremmer at all in their planning. At least in their publications there was no mention made of them. They consider their building first and foremost as an artistic-technical process.

Thus far the Zollverein has not met with fitting and critical consideration in architectural history. Some false data – in the BDA's architectural guide to Essen or in Winfried Nerdinger's book *Bauhaus-Moderne im Nationalsozialismus* – has contributed to a further misunderstanding: Zollverein XII is considered a genuine Nazi display of power. This cannot have been the case. But the regime's publications and the architects themselves contributed to this misunderstanding. The exemplary mine was quickly usurped by the Nazis and presented in books such as *Building in the New Reich* and the monthly *Deutsches*

Volkstum (German Folklore) in large photos. For the official magazine *Beauty of Work* Schupp and Kremmer themselves wrote text on the role of the architect in civil engineering. Zollverein is by no means as the architects would have had us believe – a complex which was developed purely according to function. And certainly not in a Bauhaus sense. Its axes, symmetries and monumental super-elevation served predominantly to present the United Steelworks and their subsidiary the Gelsenkirchener Bergwerks-Aktien-Gesellschaft.

The architecture of the Zollverein follows two almost opposing principles – which is what constitutes the ambivalence: monumentality and flexibility. The sheer size of the complex and the flexible building-block system of the building sheaths which can easily adjust to changing functions. Today, as the social and architectonic context which led to the building of the Zollverein XII is long since history, almost any new use is thinkable. In fact, the Zollverein was already a "symbol of structural change" when it opened back in 1932.

LITERATURE

Bund Deutscher Architekten BDA (ed.): *Architekturführer Essen* (Essen, 1982)
Burchartz, Max; Witzel, Walter (eds.): *Kohle an der Ruhr. Eine Bildfolge mit erzählendem Text* (Essen, 1932)
Busch, Wilhelm: *F. Schupp, M. Kremmer Bergbauarchitektur 1919–1974*, Landeskonservator Rheinland Arbeitsheft 13 (Cologne, 1980)
Busch, Wilhelm; Scheer, Thorsten, (eds.): *Symmetrie und Symbol. Die Industriearchitektur von Fritz Schupp und Martin Kremmer* (Cologne, 2002)
de la Chevallerie, Huberta: *Zeche Zollverein Schacht XII in Essen. Gebauter Gedanke* (Essen, 2004)
Janzen, Thomas: *Albrecht Renger-Patzsch. Photographien des Ruhrgebiets* (Ostfildern, 1996)
Nerdinger, Winfried: "Bauhaus-Architekten im 'Dritten Reich'", in: *Bauhaus-Moderne im Nationalsozalismus* ed. by W. Nerdinger (Munich, 1993)
Reger, Erik: "Die Schuldfrage der Rationalisierung", in: *Weltbühne* 1932, reprint in: Reger, E.: *Kleine Schriften*, vol. 1 (Berlin, 1993)
Schupp, Fritz; Kremmer, Martin; *Architekt gegen oder und Ingenieur*, Völter, Ernst (ed.), (Berlin, 1929)

HEINRICH BÖLL – HANS KRABEL – ACHIM PFEIFFER

RESTORATION, CONVERSION AND EXPANSION

BIGGER PICTURE

When in the early 1990s we were awarded the first contract to restore a building at the Zollverein Shaft XII we had already gained some experience in finding new uses for discontinued industrial buildings.

Ten years ago, the parish community in the neighborhood of our former office in the Altenessen sector of Essen was on the lookout for premises for a youth club and had come across the abandoned buildings of the former Carl mine. The conversion work was achieved in the style of the 1980s, namely in opposition to the officials and with much subcultural initiative and with our expert and voluntary assistance; today it houses one of the most successful cultural centers in the Ruhr District.

The contract to restore the buildings at Shaft XII arose on a completely different basis. It was not some idea spawned within the city or a particular district, it was no concrete need for space, and there was no clear notion of what the use would be. The State of North Rhine-Westphalia had placed the pits under a preservation order when they ceased operation in 1986. There was a supra-ordinated interest in maintaining the architectural remains of this discontinued industrial plant, but no intention of giving it a specific purpose.

In other words, this time the initiative to preserve the structures came from above and not from below. In the 1990s, during the “International Emscher Park Building Exhibition” we had, under the directorship of the Bauhütte Zollverein realized a whole series of conversion efforts on Shaft XII. At the end of the IBA and Bauhütte 1999 there was a brief period where nothing happened, and then the site was included in the UNESCO

Today, the architecture of Zollverein's Shaft XII is sure to attract attention. Yet for a long time, the future of the colliery was by no means sure. It was not until December 16, 1986 one week before its closure, that North Rhine-Westphalia state's application to place the whole facility under the protection of a monument was successful. Initially, there were no clear ideas as to how Shaft XII was to be used. Only when the state created the "Internationale Bauausstellung Emscher Park" (International Emscher Park Building Exhibition) (IBA) in 1989 did the first plans begin to take shape to build up the complex as a cultural center with studios, stages and a design center. In the 1990s, architecture studio Heinrich Böll/Hans Krabel carried out the first renovations of Shaft XII. The structures of Shaft XII's front part, redeveloped by the Essen-based planners, quickly became one of the highlights of IBA's 89

World Cultural Heritage (the brief covered Shaft 1/2/8 and the Zollverein coking plant) and preparation of a master plan by Rem Koolhaas, heralding a new era.

BASIC CONTEXT

The key reason for the special importance attached to the Zollverein mine complex is the architecture of Shaft XII. The architectural design of the individual buildings and their configuration to form an ensemble are outstanding examples of industrial construction.

Architects Martin Kremmer and Fritz Schupp were both in their early thirties when they were commissioned to handle the project by Gelsenkirchener Bergwerks AG. Large undertakings require courageous developers.

The Zollverein's architecture is of a quite timeless quality. However, just under twenty years ago the purpose for which they had been built ceased to exist and the mine complex was closed. The developers and their architects would have had the buildings razed to the ground as of that moment. That, after all, had been envisaged right from the beginning.

The company released the former owners of this duty and decided to preserve the shells of the buildings even though they no longer fulfilled the purpose for which they had been constructed.

New architects were employed to convert the old buildings for new purposes.

THE FAÇADES

The façade devised by Schupp and Kremmer is suspended from a load-bearing structure and is a skin that itself bears no load. It consists of a steel lattice made up of vertically and horizontally inserted rolled steel profiles that were either filled by masons using solid

projects. Since then, the architecture team has been carrying out almost all renovations and redevelopment work on the former coal mine complex and in so doing, has a number of challenges to face. Thus the boiler-house façade was so dilapidated that Böll and his colleagues had to develop a new façade with increased wind and rain protection as well as improved thermal protection. This modification to the original construction came into effect from then on for all subsequent renovations. However, Norman Foster's studio took on the transformation of the interior structure of the boiler house into a design center, while Böll's studio planned and realized the future museum building in the former coal washing building, which will first serve ENTRY2006 as an exhibition forum, together with Koolhaas' Office for Metropolitan Architecture (O.M.A.), which also developed the master plan after Zollverein

brick or had windows featuring wired glass. The beauty of the façades stems from the proportions of the steel-and-brick lattice and the materials chosen. Steel lattice façades were already a known system at the time when the original plant was erected. Schupp and Kremmer simply drew the elevations, while the details were completed by the steel engineering companies. The façades were designed to have a limited service life and the choice of structure suggested a lifespan of about forty years. After that the façades were simply replaced, which was easy as they were independent of the main load-bearing structure.

When we began our work all the façades were in need of restoration. For the former owners it would have made no business sense to prepare them for an enduring existence. The aspect of building preservation had played no role in those, the first sixty years of building and maintenance measures; we were the first to work on the façades with a view to preserving them for posterity.

The façade restoration we carried out can essentially be subdivided into two phases: the first phase of "repair work" lasted from 1989 to 1995, the second phase of "reconstruction" began in 1995. We developed the repair plans in line with monument protection criteria, and alongside the original appearance the focus was on preserving as much of the original building as possible. In other words, the existing steel-and-brick frames had to be repaired and the original bricks used wherever possible, and the existing windows simply restored. Since initial analyses showed that corrosion had impaired the load-bearing properties of parts of the steel profiles in the frames (above all the horizontal profiles, in particular the highly problematic cross-timber profiles), we set about devising additional auxiliary structures. To this end, on the inside we erected a "corset" that relieved the load on the existing structure. In other words, we did not eliminate the damage but simply stripped the damaged section of its function, meaning it could stay in

was designated as a World Cultural Heritage Site. In accordance with monument protection guidelines, Böll's studio reproduced the façades of the buildings originally developed by Fritz Schupp and Martin Kremmer so as to keep as much original substance as possible alongside the original image. Owing to the fact that parts of the steel profile had corroded and were no longer serviceable, the architects put a "corset" inside. "What results is a largely original façade with the likewise original physical and constructive weaknesses", the architects report.

place. As a result, we were able to retain most of the original façade with its likewise original structural and physical weaknesses. This will require high maintenance inputs in the future.

Starting with the conversion of the boiler house, we developed a plan for the "reconstruction of the façades". The boiler house had been out of commission for longer than the other buildings and was substantially larger. The state of the façades was comparatively poor, meaning that large sections would have been lost if we had opted even for their cautious repair. The façade of the boiler house gave rise to requirements that were greater than those for previous projects – the boiler house was designated to in future house the collection of Design Zentrum Nordrhein Westfalen.

We developed a completely new façade. We retained the main design elements of the original, namely the outer elevation of the breadths of the profiles in the frame and the choice of materials, but then devised a structure that was better in several regards. It no longer exhibits the original structural shortcomings and is therefore far more durable than the original façade. Moreover, it allows for higher-grade usage than did its industrial predecessor, as it takes into account a higher standard of impermeability to wind and rain as well as heat insulation. We achieved this goal by using changed cross-timber profiles, inserting rainwater dewatering within the façade, hydrophobic treatment of the masonry, and several other detailed measures. Since the boiler house project, we have used the modified reconstruction of the façades for all subsequent restoration work.

NEW USES

All the restoration and conversion work achieved at the Zollverein's Shaft XII to date has been handled or supported by our office. Only the design and implementation planning of the conversion of the interior of the boiler house was not, but was performed

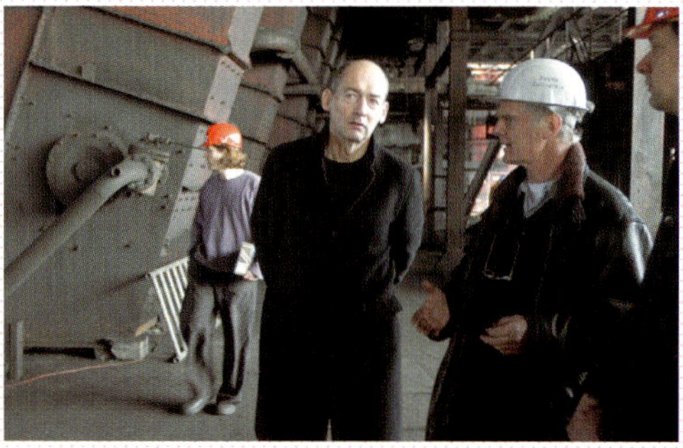

Architects Rem Koolhaas (center) and Heinrich Böll on a tour of the coal-washing building at the Zollverein mine.

instead by the Norman Foster office; we planned the work for the coal-washing plant together with the Office for Metropolitan Architecture (O.M.A.), Rotterdam.

Our task in all the projects creating new uses was initially to come up with the right architectural designs. The brief was to combine the existing building and the planned use.

With the exception of the Shaft building itself, which is still used for mining purposes, all the restored buildings at Shaft XII are no longer used for their original functions. In most cases, the structures are halls, which feature large rooms not cluttered by supporting pillars and not assigned to specific types of use.

The first to be converted was Hall 5, a hall that housed workshops and which we restored for usage, without making any changes to the layout. documenta 9 used the hall in 1992 as an exhibition venue for a piece by Ulrich Rückriem. He found it quite natural to position his piece in the hall, showing right at the beginning of the restoration work at the Zollverein how universal the buildings there are.

Our work does not aspire to preserve some historical patina, to create an ambience suffused with industrial romanticism. We restored the stringency and clarity of the hall that Schupp and Kremmer had originally designed. It constitues a prototype for the edifices at Shaft XII; all the other buildings are essentially variants on this theme.

In all subsequent restoration work we continued to prioritize identifying existing spatial qualities and optimizing them for future uses. In part, the new uses required the existing building volumes to be extended. Thus, Hall 9 – the new canteen – required an annex for the kitchen and a lobby; and the boiler house was given an additional freight elevator on the North side. These additions always take back seat to the historical structures. One departure from this approach was the gangway to the coal-washing plant designed together with O.M.A. This comparatively high-profile gesture points to the

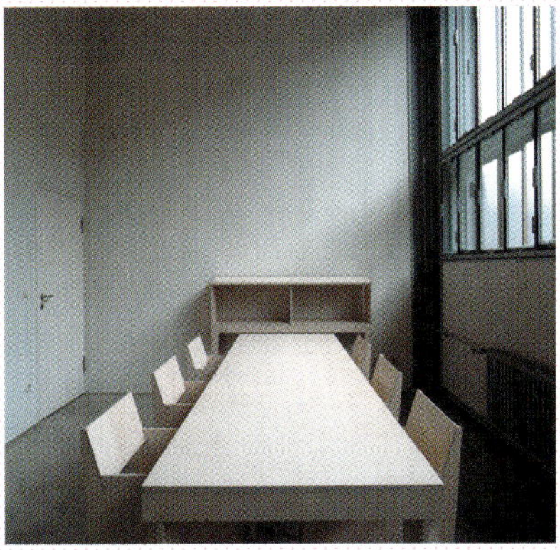

Interior of the reconstructed Hall 6, 1993

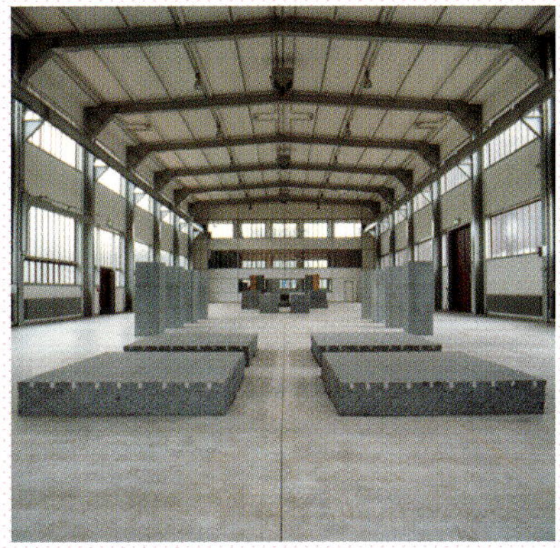

View into the reconstructed Hall 5, 1992, Ulrich Rückriem exhibition

new beginning and the opening of the Zollverein. It is a gesture that must remain unique at Shaft XII.

Diverse uses have hitherto been enabled: a restaurant in Hall 9, an ad agency in Hall 2, workshops and offices in Hall 10, studios and an events auditorium in Hall 12, an art gallery in Hall 6, a museum in the coal-washing plant. New uses are continuing to evolve, and Hall 6 has been converted from an exhibition house into an administrative building without any fundamental changes being needed.

CONCLUDING REMARKS

In the long term, the Zollverein buildings cannot be discontinued monuments to the past. For their built substance will not allow that to happen, as it simply decays.

After over fifteen years of building work at Zollverein, visitors still sometimes ask us: "What did you actually do here?" – We have through our efforts enabled new uses in existing buildings. And in the process the latter have retained their original character. The intentions behind Schupp and Kremmer's blueprints are fully discernible.

ANDREJ KUPETZ

TRANSFORMATIONS

If you were to imagine an ideal location for illustrating a region's transformation from a site of industrial production to a forward-looking center for innovative services, a place where in a climate of ongoing inspiration creative industries can thrive and innovative product and service concepts can develop, Zollverein would hardly be the first place to spring to mind. But then again why not?

Admittedly, at first sight the Ruhr District only world cultural heritage site would seem to be the exact opposite of a center of creativity. Seldom has the word industrial culture been applied to as fitting a place as Zollverein. After all, it symbolizes nothing less than the rise and fall of a region's industrial age. The former coal mine bears witness to the powerful rise of coal-mining in the Ruhr District since 1850, which in the 1920s was expressed by an architecture of machine aesthetic brimming with self-confidence, and that claimed to epitomize a new age and a new town. And it relates the process of decline in a region, the industries that characterized it and ultimately its symbols. Yet Zollverein has a charm all of its own. It exudes a very special fascination arising from the fact that it documents change, decline and revival, but also from the solid presence of the bygone industrial age contrasted with pioneering utilization concepts – two phenomena that initially seem totally irreconcilable.

The structural change in the Ruhr District, which we experience today as the gradual decline of a region's traditional industries with all the economic and social repercussions involved, first began over forty years ago. Indeed, perhaps we must concede that the region's story is essentially an ongoing story about the district's changing economic and social structures of which the Ruhr District with its coal-mining, iron and steel in-

From the buildings via the plant and machinery to the street lamp, the logic behind the design of the Zollverein coal mine is second to none. The author sees it as lucky for its future "that we postindustrial people are rediscovering the romanticism of the mechanical age, a time that smells of machine oil and tastes of coal dust". According to him, the extraordinary quality of experience of the spacious complex corresponds to the desire for an authentic experience. His story ends neither with the decline of the industrial age in the Ruhr District nor with the designation of the mine as a UNESCO World Cultural Heritage Site. Alongside tourist attractions, creative companies and platforms for the debate on design, such as the Red Dot Design Museum and the Zollverein School of Management, have moved to Zollverein. Kupetz rejects as unfounded objections that the central thematic thread is

dustries, with Krupp steel and coal-miners choirs forms only one chapter. And perhaps it is precisely the seemingly irreconcilable nature of the dominating industrial architecture with new forms of usage and the resulting decision not to preserve the Zollverein solely as a museum complex but to intervene in the existing structures and develop additional areas that serve to forge a special link between past and present. Now at least there is a chance that the history of Zollverein need not end with the last ton of coal mined there. Nor does its history end with the mere conservation of the development, as was expressed in Norman Foster's reticent approach to that part of the building, which today houses the red dot design museum. And it does not end with its being declared a World Cultural Heritage Site by UNESCO. Zollverein need not be the end. Rather, it has a future, if in addition to being an attraction as a museum the history of this location continues to be told.

Seen against this background, commissioning Rem Koolhaas to develop a master plan for the Zollverein complex is part of an overall forward-looking strategy. It would be wrong to interpret the choice of Koolhaas as another case of using great architects as skilled ambassadors of change and their architecture as a means of dissemination. Local politicians from all over the world are familiar with the so-called Bilbao effect, which succeeded in lending the small town of Bilbao international importance and attracting streams of tourists interested in art to the Basque province. Emulators of these post-industrial utilization strategies for industrial complexes are to be found everywhere in the world where the industrial age has already taken its leave or is about to draw its final breath. In England's poorly developed North, just as much as in Herford or Wolfsburg, where Iraqi architect Zaha Hadid was able to realize her own Bilbao effect. From Wolfsburg's main station to the entrance to the VW AutoCity there extends a futuristic block

somewhat strained. He argues that the industrial age brought forth this specific form, and in the same way gave rise to design as a discipline, so why should not Zollverein also become the location of a new design concept, asks Andrej Kupetz and remembers the spatial and application concept of renowned architect Rem Koolhaas, which the latter developed in the form of a master plan. As an author, Koolhaas left his mark on the term "junk space" in terms of the development of large cities going off the rails, in which urban planning is no longer possible as the planners are only concerned with serving the diverging and selfish interests of its users. In his publications, Koolhaas brought, among other things, the outskirts of cities back into the architectural eye. The spatial situation of the Zollverein colliery is no different. When people were concerned with separating work and play in the urban scene and the old colliery

whose appearance reminds you of a spaceship from an alien galaxy that went out of control and happened to crash over Wolfsburg. It is a work with which Zaha Hadid has left her very own mark on Wolfsburg.

It is a different situation with the Zollverein. It was a stroke of genius not to commission Koolhaas to design a new building, but rather to draw on the architect's communicative strength in developing a space and utilization concept. For years Koolhaas earned a reputation writing about architecture, without working as an architect. With his theories he penetrated deeper into the current problems of our society and its buildings than a great many of his colleagues. He coined the term "junk space", which perfectly described the sprawling development of cities in which urban planning was no longer possible because the planners were much too preoccupied with slavishly satisfying the highly divergent and selfish interests of their users. By contrast, Koolhaas protested about a rampant mentality of prettifying ugly places and returned cities' outlying peripheries to the focus of architectural interest. Zollverein had suffered the fate of other peripheral developments. Located at the north-eastern outskirts of Essen, the complex increasingly became a kind of *persona non grata* in the city's post-war urban planning. When it became chic to also separate work and leisure from one another in the urban landscape and the old coal mine was no longer viewed as a symbol of work, the pride of the town, it instead became an oversized non-place whose pervasive and highly visible architecture merely reminded people that here an age would soon come to an end that would leave the city and region alike a swathe of unresolved problems as a legacy.

Zollverein was lucky enough to benefit from the change in societal values; specifically, the more we post-industrial people are liberated from physical work and industrial production processes, the more we discover the romantic nature of the mechanical age – an age that smells of engine oil and tastes of coal dust. While we are happy enough at

was no longer seen as the pride of the city, it degenerated into an oversized negative area. Koolhaas' master plan, the renovation of the former coal-washing building, which his OMA saw through together with the Heinrich Böll architecture studio, and the architecture of the Japanese group Sanaa for the Zollverein School have made Zollverein into a model for a successful process of transformation in terms of work, culture, architecture and design. For its future development, it is important to acknowledge that it is not finished, but is very much a work in progress.

›
New and refurbished Buildings: Zollverein School of Management and Design, SANAA, Sejima and Nishizawa architects, Tokyo (left and right above), the coal-washing building (left and right below), location of ENTRY- 2006 and upcoming new Ruhrmuseum, OMA Rem Koolhaas Rotterdam. Both architects worked in cooperation with architects Boell/Krabel, Essen.

having escaped the physical grind of the industrial age, we simultaneously long for the physical quality we ascribe to this era namely of being able to experience our environment authentically. The fact that we see former sites of industrial production as places to experience industrial culture, cultivate them and promote them as tourist attractions can be attributed to this desire for authentic experiences. And Zollverein offers this very kind of experience in a unique, unmistakable manner: Nowhere else is there a comparable industrial complex, which was planned and realized as a cohesive whole. The old Zollverein can justly be described as a total artwork of the industrial age – rigorously designed from the streetlamps via the buildings to machinery and unparalleled the world over.

Consequently, if alongside the tourist development of living industrial worlds, the new creative industries make Zollverein their home and continue the story of the region's industry, this is not a matter of implanting alien elements onto historical ones; rather you can say the liberation from the dogma of industrial design that form should follow function is presented in a three-dimensional form. For instance, an Apple iPod could take on any external shape, since its control chip does not give any information about which shape would be suitable. Yet designers gave it a shape that recalls the electrical gadgets of the 1960s. Why then should not the creative industries develop products and concepts for the post-industrial world inside the former shells of the industrial age? Though the comparison might seem far-fetched, both the iPod and the new users of the Zollverein mine avail themselves of industrial modernism's design language. And the transformation of the contents is possible because the external form has infinite quality of experience to offer its users.

As such the skepticism voiced fairly frequently to the effect that design were merely an artificially construed means of giving the complex a new function is unfounded. In the

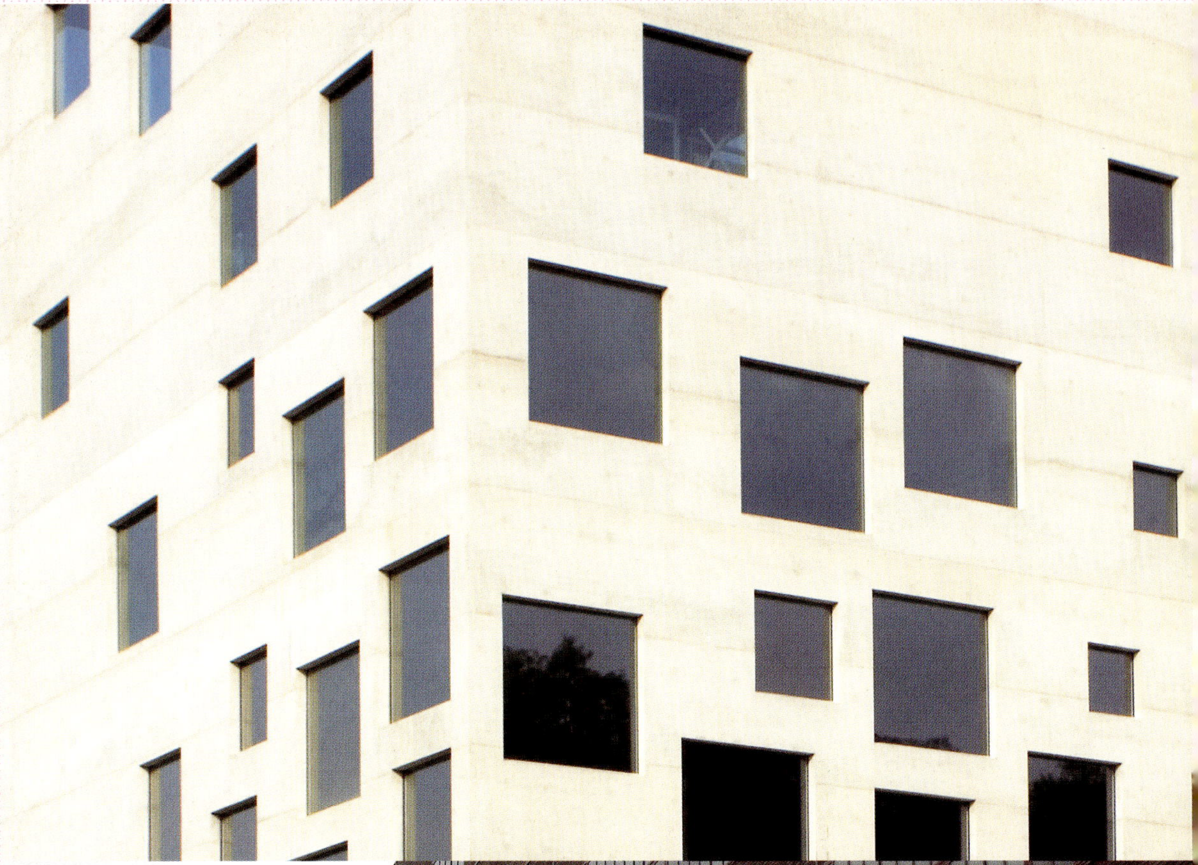

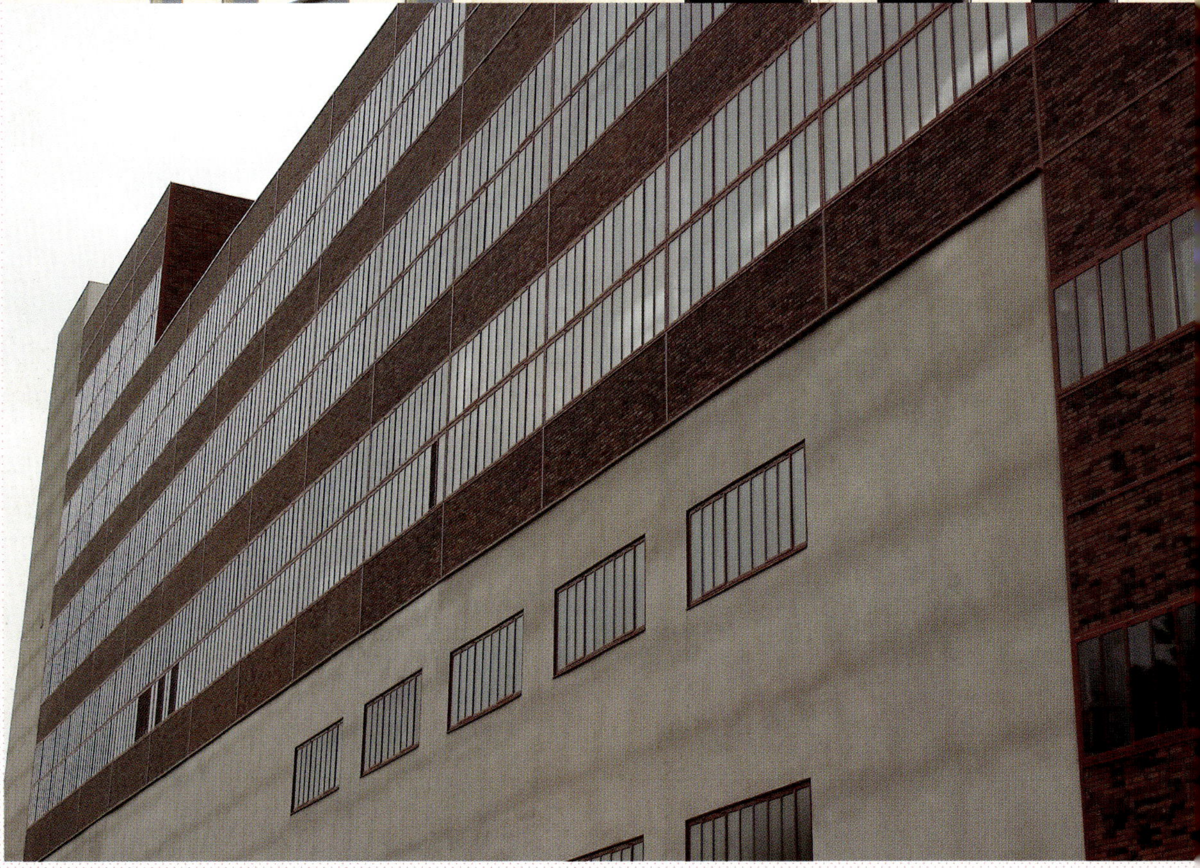

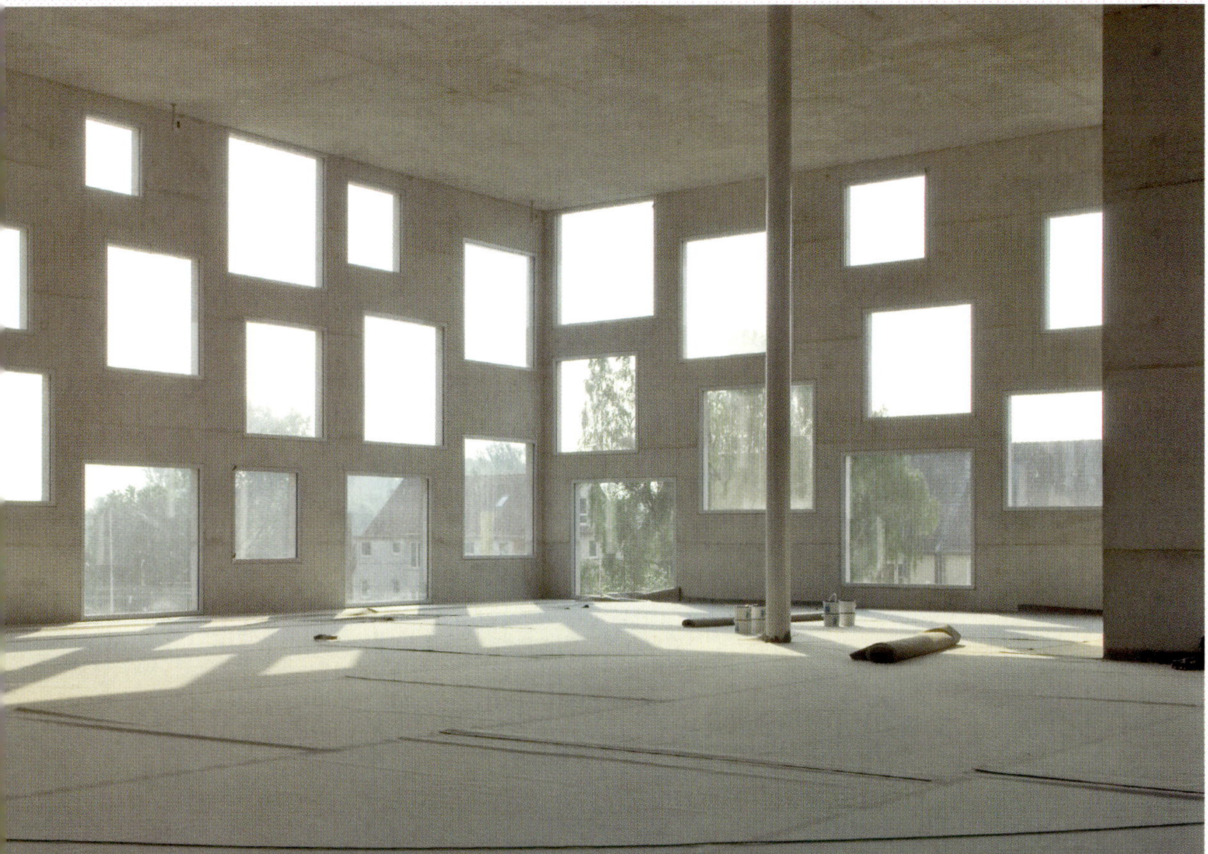
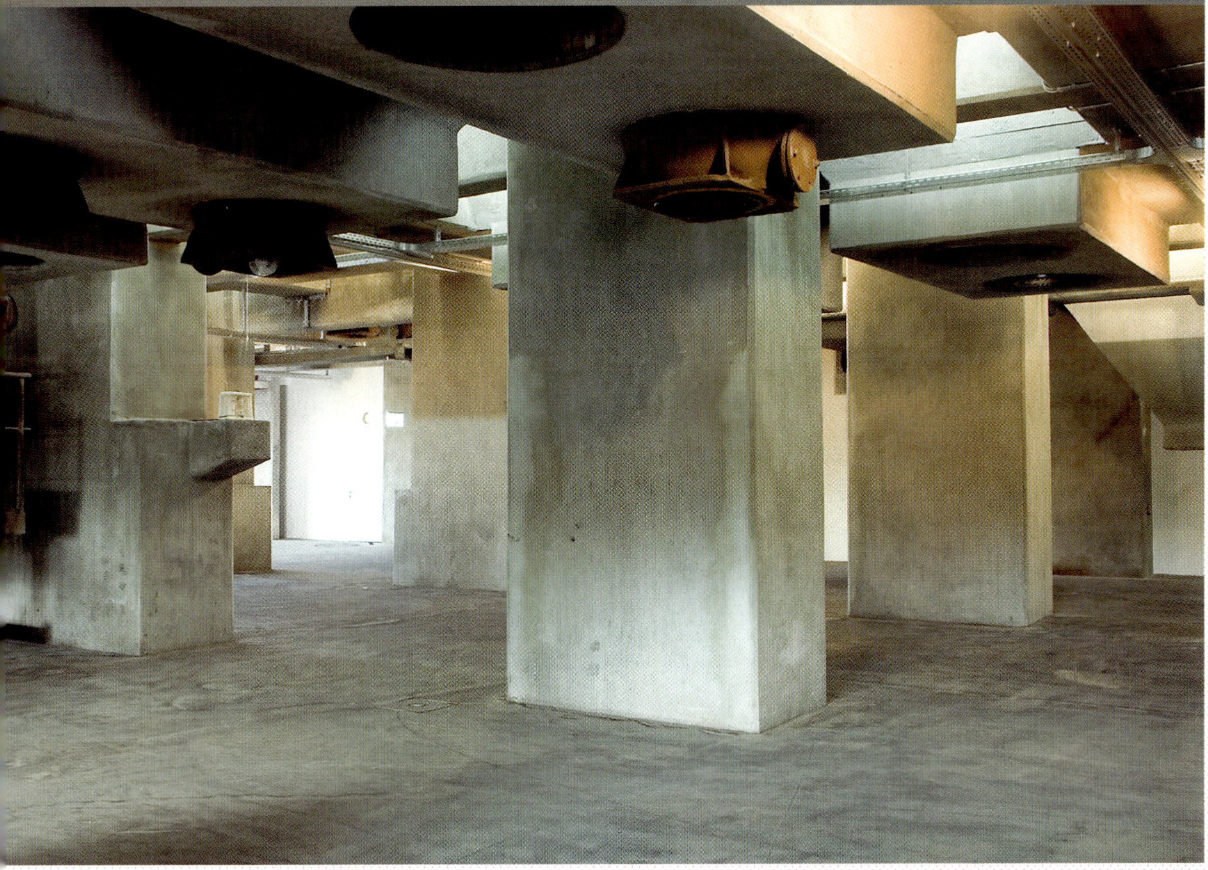

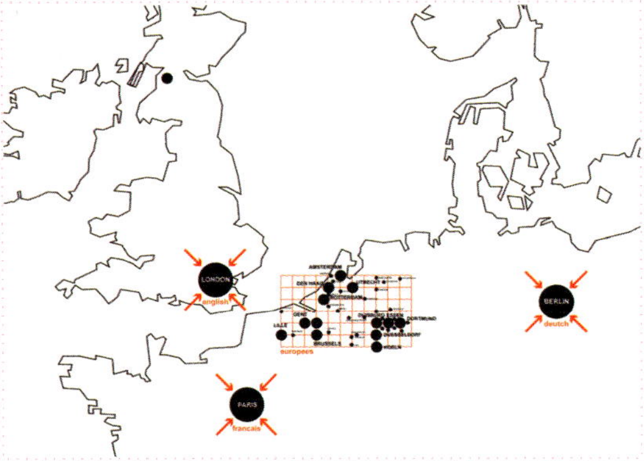

The Office for Metropolitan Architecture's (OMA) master plan stresses the extensive and at the same time dense infrastructure of the Ruhr District in the context of major European cities.

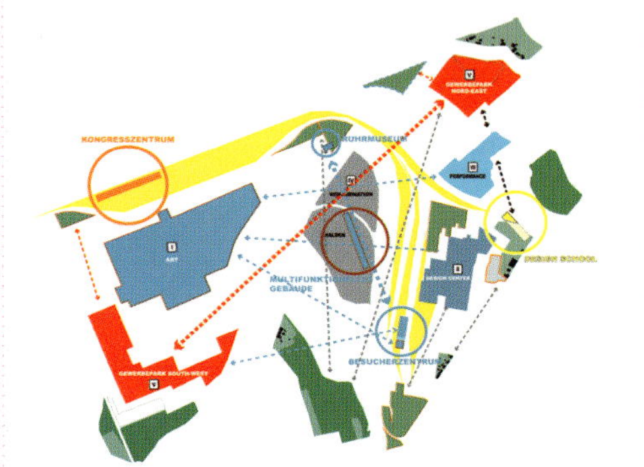

In its master plan, the OMA places an outer ring of attractions around the existing complex, in which a series of new buildings is to be concentrated.

same way that we increasingly remove design from its industrial context and depending on our perspective grasp design as a service for marketing or as motor of economic innovation, Zollverein has also been removed from its industrial context. The industrial age produced Zollverein and design as a discipline, why should it not be possible to also view the "new Zollverein" as the location for a new understanding of design?

The proposed space and utilization concept will enable new design to evolve in the "World Cultural Heritage Site of design". This will be made possible not only by the creative industries, which are to form a business cluster on the Zollverein complex, but also the creation of international platforms for an interdisciplinary discussion on design quality, design training and design innovation. The Red Dot Design Museum is an effective measure of current international design, while the Zollverein School of Management and Design is unique in its networking of business and design. The spectacular architecture of this design institute, which was developed by the Japanese group Sanaa, will attract international interest. The former coal-washing plant, which Rem Koolhaas' office OMA together with architect Heinrich Böll converted into an exhibition building, will in future house the Ruhr Museum and initially ENTRY2006, which as an explorative design exhibition attempts to create a tableau of the current and future issues of this discipline in the context of business, science and society.

Zollverein continues to evolve. Today, this World Cultural Heritage Site is already a tourist attraction drawing over half a million visitors a year. And the eighty firms located here, many of them from the area of creative industries, provide around 1,000 jobs. If you consider that thirty years ago during the final golden age of coal mining in Essen, 5,000 people were employed at Zollverein, the current number of jobs represents a considerable success. As such, Zollverein is also exemplary for successful transforma-

Foster and Partners designed the internal structure of the former boiler house at the Zollverein coal mine. The red dot design museum there offers a highly exciting contrast between the industrial aesthetics of the early 20th century and contemporary design.

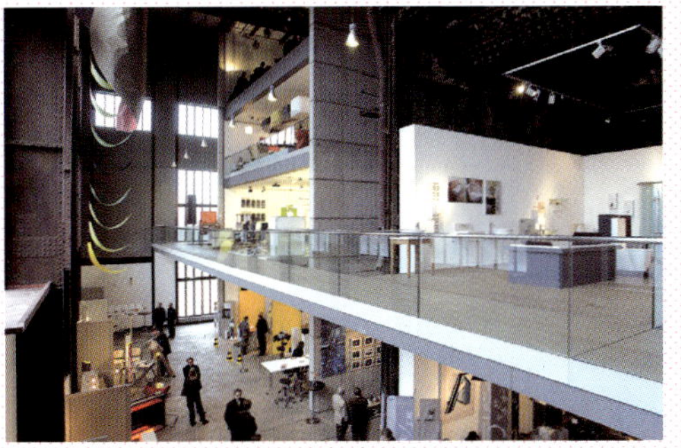

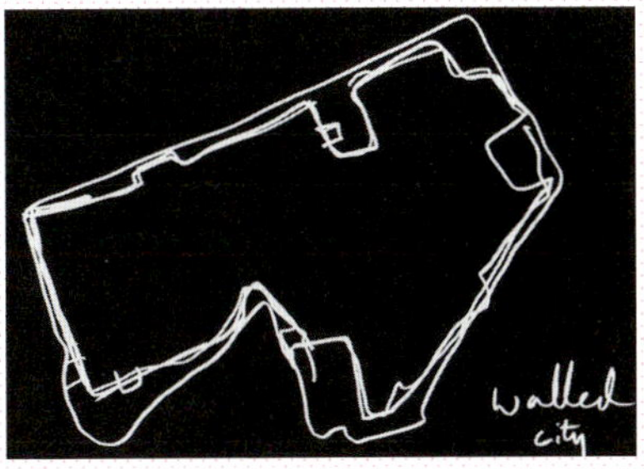

Rem Koolhaas sees the central complex of the old Zollverein mine as a " walled city ", while at the same time the strengthening of the peripheral zone should facilitate a new interpretation of the existing elements.

tion, specifically the transformation of work, culture, architecture and design. For the future of Zollverein it will be important to recognize that this process of transformation is not complete but in the best sense of the word represents "work in progress".

AUTHORS' BIOGRAPHIES

Heinrich Böll Born in 1940 in Essen, Böll is an architect. He studied Architecture at Berlin's Technical University (1961–1968) and has worked as a self-employed architect in Essen since that time. Since 1975 he has collaborated on an ongoing basis with fellow architect Hans Krabel. Böll has been a member of DWB, Deutscher Werkbund since 1995 and has participated in international competitions.

Norbert Bolz Born in 1953, Bolz is a media scholar and essayist. He read Philosophy, German and English Literature, and Religious Studies in Mannheim, Heidelberg and Berlin and was awarded a Ph.D. for a thesis on Adorno's aesthetics supervised by philosopher of religion Jacob Taubes. His habilitation was on "Philosophical Extremism between the World Wars". Bolz was Taubes' assistant until the latter's death. 1992–2002 Professor of Communications Theory at the Institute of Art and Design Studies at the Universität GH Essen. Since 2002 /03, Professor at the Berlin Technical University, Faculty I Humanities, Institute for Language and Communications, Dept. of Media Studies. Norbert Bolz has edited countless publications on media history, the theory of the mass media and communications theory and network logic. In recent years, among other things, he has published: *Weltkommunikation* (Munich, 2001); *Das konsumistische Manifest* (Munich, 2002); *Blindflug mit Zuschauer* (Munich, 2005); *Die Helden der Familie* (Munich, 2006).

Roger M. Buergel Born in 1962, Buergel is an exhibition curator and author, and has two children.
He studied Art and Philosophy in Vienna. Curated shows include: *Dinge, die wir nicht verstehen* (with Ruth Noack, Generali Foundation, Vienna, 2000); *Gouvernementalität. Kunst in Auseinandersetzung mit der internationalen Hyperbourgeoisie und dem nationalen Kleinbürgertum* (Old Kestner Society, Hanover, 2000), *The Subject and Power – the lyrical voice* (CHA Moscow, 2001); *Die Regierung* (with Ruth Noack, Art Hall of the University of Lüneburg; MACBA – Museu d'Art Contemporani de Barcelona; Miami Art Central; Secession, Vienna; Witte de With, Rotterdam, 2003–05). Artistic Director of documenta 12 (2007).

Heike Edelmann Born in 1964 in Essen, Edelmann is a journalist and copy-writer; she lives in Hamburg, is married to Thomas Edelmann and has one son. From 1988–1990 she worked for the magazine *Wolkenkratzer Art Journal*. She graduated in Art Studies in Frankfurt. Heike Edelmann is in charge of the ranking of European agencies for corporate design – initially for *Horizont* magazine and then later for *Page*. She freelances for Deutsche Presseagentur (dpa) on design topics.

Thomas Edelmann Born in 1963 in Frankfurt, Edelmann lives in Hamburg. Edelmann is a journalist and primarily writes on design and architecture. He studied Law in Frankfurt/Main in the early 1980s and from 1987 helped set up the magazine *Design Report*, which he accompanied when it moved to Hamburg in 1993; he was editor in chief from 1996–2001. Since then he has worked freelance as a journalist and publicist, among others for *A&W Architektur & Wohnen, form – The Making of Design* and *design report* and Austrian daily *Der Standard*. Edelmann edited the book *Tara – Armatur und Archetypus* (Basel, 2003).

Frank Edenhofer Born in 1968, Edenhofer is a biochemist and a biologist specialized in stem cells, and has three children.
He studied Chemistry at the University of Munich (1989–1994) and was awarded a Ph.D. in Biochemistry for a thesis on infectious protein particles. He went on to work at the University of Cologne with Klaus Rajewsky, a pioneer in the deliberate alteration of genes in mice. In 2002 winner of the New Talent Research Prize for the North-Rhine Westphalia Competence Network Stem Cell Research, worth EUR 1.25 million. He is currently lecturing and doing research at the University of Bonn, where he is focusing on questions of stem cell differentiation for cell regeneration in the nervous system. Countless publications and lectures at national and international conferences on issues relating to targeted mutagenesis and stem cell differentiation.

Jochen Eisenbrand Born in 1970, Eisenbrand did a masters degree in Applied Cultural Studies at the University of Lüneburg and has worked since 1998 for Vitra Design Museum. As a curator he has prepared various exhibitions there that have traveled from Weil am Rhein on to other museums in Europe, the United States and Japan. They have included an Isamu Noguchi retrospective, a show on Ingo Maurer and most recently *Airworld – Design und Architektur für die Flugreise*. Jochen Eisenbrand is Co-curator of the exhibition *Open House: Architecture and Technology for Intelligent Living*, which was organized in collaboration with the Art Center College of Design in Pasadena, California.

Francesca Ferguson Born in 1967, Ferguson studied at Somerville College, Oxford University, graduating in 1988. In 1989 she worked in Berlin as Associate Producer at ABC News Network. In 1992 she started as a freelance correspondent and production assistant in Asia, the Soviet Union and East Europe, among others for Channel 4, TF 1, BBC 2, and Spiegel TV. She then worked as an assistant curator at Kunst-Werke Berlin and on co-organizing "Club-Berlin" at the Venice Biennial. In 2004 she curated *Deutschlandscape – Epicentres at the Periphery* for the German pavilion at the 9th Architecture Biennale of Venice. She organizes the gallery program at Akademie Schloss Solitude in Stuttgart and initiated and chairs urban drift, an international network on architecture and urbanism; since autumn 2006 she is director of the Basel Architecture Museum.

Vilém Flusser Born in 1920 in Prague, Flusser started reading Philosophy there in 1938. In 1939, he fled the Nazis and lived until 1940 in London, emigrating after graduation to São Paulo, together with his wife Edith. Until about 1950 he worked there in the import/export business. In 1950–51, he participated in a book project on the intellectual history of the 18th century and in 1959 was appointed Lecturer in the Philosophy of Science and in 1963 Professor of Communications Philosophy at the University of São Paulo. In 1972, Flusser fled the military regime in Brazil and settled in Southern France. In 1991, he was a guest professor at the University of the Ruhr, Bochum. He was killed that year in a car crash. Selected works: *Bodenlos. Eine philosophische Autobiographie* (Cologne, 1992), *The Shape of Things. A Philosophy of Design* (Göttingen, 1993) *Gesten. Versuch eine Phänomenologie* (Cologne, 1996) *Vogelflüge* (Munich, 2000)

Hans Krabel Born in Berlin in 1940, Krabel is an architect and town planner. Krabel studied Ehtnology and Art at the University of Munich (1960–1961) and then at Berlin Technical University (1962–1968). From 1968–1971 he worked in various planning offices in Düsseldorf and Essen. From 1972–1974 he took teacher training in Munich and also passed the Major State Examinations in Urban Planning. He has collaborated on an ongoing basis with architect Heinrich Böll since 1975. Since 1979 he has been an active member of the "Initiative Zentrum Zeche Carl e.V.".

Andrej Kupetz Born in 1968, Kupetz is a design manager and lives in Frankfurt/Main and has a child. Since 1999, Kupetz has been Managing Director of the German Design Council, Frankfurt/Main. Since August 2006 he is President of Zollverein School of Management and Design in Essen.
He studied Industrial Design, Philosophy and Product Marketing in Berlin, London and Paris. After freelancing as a design manager he was Head of Section at the Design Transfer Institute at the Berlin University of the Arts, moving in 1997 to Deutsche Bahn AG. There he was in charge of Group brand management and of the implementation of various corporate design processes. Kupetz publishes regularly on design-related themes in specialist publications such as *Design Report*, *form* and *Horizont*.

Werner Lippert Born in 1950, Lippert is a corporate culture consultant. He studied Medicine and Art History in Aachen and Bonn. At the beginning of the 1970s, he was active as an art reviewer and exhibition curator, later moving into advertising as a copy-writer, Creative Director, and Managing Director; in 1987 he founded his own Düsseldorf-based agency Lippert, Wilkens Partner. Since 1996 he has joined forces with Petra Wenzel to form the consultancy company "Projects – Corporate Culture Consultants", in Düsseldorf. Lippert is in charge of exhibition management at NRW-Forum Kultur und Wirtschaft, which has included shows such as *Mutanten – die deutschsprachige Comic Avantgarde der 90er Jahre* (1999), *Die Kunst zu Fliegen – in Fotografie und Film* (2004), *Vivienne Westwood* (2006). He is a specialist author on topics relating to culture and advertising, and edits the *Annual Multimedia*.

Ellen Lupton Is an author, curator, blogger and graphic artist. She runs the MFA Program for Graphic Design am Maryland Institute College of Art (MICA) in Baltimore. She is curator of contemporary design at the Cooper-Hewitt National Design Museum in New York City, where she has been responsible for numerous exhibitions and publications. Examples: *the National Design Triennial Series* (2000 and 2003), *Skin: Surface, Substance + Design* (2002), *Graphic Design in the Mechanical Age* (1999), *Mixing Messages* (1996), *and Mechanical Brides: Women and Machines from Home to Office*. She has authored: *Thinking with Type: A Critical Guide for Designers, Writers, Editors, and Students* (2004). In January 2006 together with students from the MICA graduate program she completed her book *D.I.Y: Design It Yourself*. Her career has focused on promoting a critical public awareness of graphic design. A graduate of the Cooper Union (BFA, 1985) she is currently writing a dissertation on communications design, to be submitted at the University of Baltimore.

Koert van Mensvoort Born in 1974, van Mensvoort is Co-Director of the All-Media-Foundation, an organization that visualizes contemporary cultural issues. Van Mensvoort lives in Amsterdam, having studied Philosophy at the University of Utrecht. In 1997, he was awarded an M. Sc. in Computer Science from the Technical University of Eindhoven and in 2000 a Master of Fine Arts from the Sandberg Institute in Amsterdam, where he now teaches. In addition, he is Assistant Professor in the Industrial Design Dept. of the Technical University of Eindhoven. Together with Michiel Schwarz and Mike Gerritzen, in 2005 van Mensvoort edited the book *Next Nature*, brought out by Bis Publishers, Amsterdam.

Wolfgang Pauser Born in 1959, Pauser spends his time in Vienna half writing essays on everyday culture, design, architecture and the applied arts (www.pauser. cc) and half advising companies in the field of brand and design development through his company Different Futures (www.differentfutures.com). A Doctor of Law, he worked as an art critic for Austrian daily *Der Standard* and for the German weekly newspaper *Die Zeit*, taught at the University for the Applied Arts, Vienna, and curated exhibitions (e.g., *Einbild – Wiener Secession*; and co-curated *Wunderblock – eine Geschichte der modernen Seele* – Vienna Festival). Books: *Dr. Pausers Werbebewußtsein* (Brandstätter, 1995), *Dr. Pausers Autozubehör* (Hanser, 1999).

Achim Pfeiffer Born in 1969 in Kaufbeuren/Allgäu, Pfeiffer is an architect. After graduating from the Rhenish Westphalian Technical University in Aachen in 1997, he started working for the Heinrich Böll office. Achim Pfeiffer lives in Essen.

Ulf Poschardt Born in 1967, Poschardt is a book author and media consultant. He lives in Berlin. He has written *DJ Culture*, a cultural history of disc jockeys from 1904 to the present (1995). In his other books, Poschardt has focused on fashion (*Anpassen*, 1998), coolness (*Cool*, 2000) and speed (*Über Sportwagen*, 2002) and currently on loneliness *Einsamkeit – die Entdeckung eines Lebensgefühls* (2006). Poschardt was editor-in-chief of the supplement of the *Süddeutsche Zeitung* from 1996 to 2000 and was one of the senior editors at *Welt am Sonntag*. He is currently in charge of the editorial development team for the German version of *Vanity Fair*.

Peter Reed Born in 1955, with a Ph.D. in art history, Reed is Senior Deputy Director for Curatorial Affairs at the Museum of Modern Art, New York, where he has worked since 1992. In 2005 he curated the first survey there on landscape design, entitled *Groundswell: Constructing the Contemporary Landscape*. Other exhibitions have included: *AUTObodies: speed, sport, transport* (2002) and *Alvar Aalto: Between Humanism and Materialism* (1998). From 1989–1992 he taught Art History at the University of Pennsylvania where he was also Director of Research for the Architectural Archives and the Louis I. Kahn Retrospective Project. His publications include catalogs on the shows *Groundswell* and *Alvar Aalto*. He is co-editor of *The Show to End All Shows: Frank Lloyd Wright and The Museum of Modern Art, 1940* (2004) and the catalog of the exhibition *Frank Lloyd Wright: Architect* (1994), the most comprehensive retrospective of Wright's work since his death in 1959. Reed has contributed to numerous publications, including *World War II and the American Dream* (National Building Museum and MIT Press, 1995) and *Louis I. Kahn: In the Realm of Architecture* (Los Angeles Museum of Contemporary Art, 1991).

Jan van Rossem Born in 1961, van Rossem is Senior Editor at the journal *A&W Architektur & Wohnen* and responsible for Design, Travel and Lifestyle. Each year, he portrays the *A&W Designers of the Year*, and is also in charge of organizing the prize-giving ceremony and accompanying show of what is one of Germany's most important and best-paying design prizes – and is constantly discovering new, young and creative minds, whom he cameos in *Architektur & Wohnen*.

Gerhard Seltmann Born in 1954, Seltmann lives in Flechtingen. He is Managing Director of the Ausstellungsgesellschaft Zollverein (AGZ) exhibition company and organizer of ENTRY2006. Seltmann graduated in Communications Studies in Münster. From 1989–1995 he was Deputy MD of IBA Emscher-Park, and from 1995–2000 MD of Expo 2000 Sachsen-Anhalt GmbH. From 2001 to 2002 he was in charge of the Regional Development, Foreign Relations and Tourism section at the Economics Ministry of North Rhine Westphalia. In Flechtingen, he runs GseProjekte, a consultancy for regional development, culture and tourism projects.

Peter Sloterdijk Born in 1947 in Karlsruhe, Sloterdijk is a cultural scholar, philosopher, publicist, author and essayist. He studied Philosophy, German and History at the University of Munich (1968–1974) and was awarded a Ph.D. in Hamburg in 1975 with a study on the philosophy and history of modern autobiographical literature. Since 1980 he has worked as a freelance author, publishing countless pieces on contemporary issues, on the philosophy of culture and religion, on art theory, and psychology. In 1992, Sloterdijk was appointed Professor of Philosophy and Media Theory at the Karlsruhe Academy of Design. In 1993, he was made Dean of the Institute for the Philosophy of Culture at the Academy of the Applied Arts in Vienna. Since 2001 he has been Rector of the Karlsruhe Academy of Design. Moreover,

he has held visiting professorships at Bard College, New York, Collège International de Philosophie, Paris and at the Eidgenössische Technische Hochschule, Zurich. Sloterdijk is anchorman of the ZDF show "Das Philosophische Quartett."
Publications in recent years: *Nicht gerettet. Versuche nach Heidegger* (Frankfurt/Main, 2001); *Luftbeben. An den Quellen des Terrors* (Frankfurt/Main, 2002); *Sphären III – Schäume, Pluralistische Sphärologie* (Frankfurt/Main, 2002); *Im Weltinnenraum des Kapitals* (Frankfurt/Main, 2005).

Alexander von Vegesack Born in 1945, von Vegesack is a freelance curator and collector of industrial furniture design, in which capacity he established the Thonet Museum in Boppard am Rhein and advised international museums on their collections and purchases for their design collections. He has organized shows in Paris for Centre Georges Pompidou and for Musée d'Orsay and devised the first major show on the history of industrial furniture design for the American Federation of Arts – it went on display in ten leading US museums. In 1989, together with furniture makers Vitra he founded the Vitra Design Museum, of which he has since been Director. He developed the concept for an independent foundation that is almost exclusively financed by its own output. Von Vegesack is editor of countless publications and initiates an annual international museum conference at which the key European museums exchange ideas and programs. For over a decade, he has together with Vitra Design Museum and Centre Georges Pompidou organized an internationally recognized seminar program in France, in which designers and architects as well as artists, craftsmen and skilled artistic craftsmen from a variety of different fields take part. Alexander von Vegesack was a member of Fonds National d'Art Contemporain, is a Chevalier de l'Ordre des Arts et des Lettres and has been awarded the Cross of the Order of Merit of the Federal Republic of Germany.

Peter Wippermann Born in 1949, Wippermann lives in Hamburg, is married and has a daughter. Wippermann is the founder and partner of Trendbüro – Beratungsunternehmen für gesellschaftlichen Wandel GmbH, and since 1993 has been Professor of Communications Design at the Duisburg-Essen University. He is also a member of the Advisory Councils of bayerndesign and hamburgunddesign.
Peter Wippermann initially worked as Art Director at Rowohlt-Verlag and at *ZEITmagazin*, before in 1988 joining up with Jürgen Kaffer to found Büro Hamburg, now one of Germany's most renowned graphic agencies. In 1990, he developed the future events *Talk with Tomorrow* for Philip Morris and was the chief editor of the future magazine *Übermorgen*. In 1992 he founded Trendbüro, and in 1993 was appointed to the Chair of Comunications Design at the University of Essen. In 2002, he co-founded the LeadAcademy für Mediendesign und Medienmarketing.

Fabian Wurm Born in 1957 in Essen, Wurm studied German Literature and Sociology at the University of Frankfurt. From 1987–1992 he was chief editor of *design report* in Frankfurt/Main, and from 1993–1998 editor of *form – Zeitschrift für Gestaltung*. In 1999, he was head of PR at the Rempen & Partner ad agency in Düsseldorf. From 2000–2003 he worked as editor for the newspaper *Horizont* and also lectured on the history of design at Stuttgart's State Academy of the Applied Arts. Since 2004 he has freelanced as a journalist, editor and author on design and architecture.

ILLUSTRATION CREDITS

--- A ---

Azra Aksamija, p. 180
Peter Allen and Carla Ross Allen, Brooklyn, USA, p. 125, left
Joe Arcidiacano, USA, p. 132, left
Architecture Research Office LLC, New York, NY, p. 134
Atelier 17/Christa Panick, p. 191
Anthony Aziz+Tony Cucher, New York City, USA, p. 130–131

--- B ---

Belux, p. 126
Corné Bastiaansen, p. 135
Uwe Bergk, Courtesy Panelite, US, New York and Los Angeles, USA, p. 133, left
Hieronymus Bosch (1450–1516), The Garden of Earthly Delights: Hell (right wing of the triptych), c. 1500 (oil on panel), detail, Prado, Madrid, Spain/The Bridgeman Art Library, p. 79
Richard Brine, p. 57
Büro Hamburg, p. 27, left
Susanne Brügger, p. 215
William J. Buchser, USA, cover
Max Burchartz, ©2006, ProLitteris, Zurich, p. 206, left
Mark Burry, p. 144

--- C ---

Jean François-Carly, p. 56
Elinor Carucci, New York, USA, p. 125, right
Nuno Cera/Talking Cities, urban drift, p. 172–173
Michel Cluizel, p. 73, 74
Cold Spring Harbor Laboratory Archive, p. 30
Computer History Museum, California, p. 27, right

--- D ---

DaimlerChrysler, p. 65
Thomas Duval. p. 11

--- E ---

Jason Evans, p. 152

--- F ---

Field Operations, p. 195
Chrystelle Fontaine, Julia Friedman Gallery, p. 16, left
Lavinia Fontana (1522–1614), Portrait of a Girl Covered in Hair, Private Collection/ © Bonhams, London, UK/The Bridgeman Art Library, p. 80, left
Forschungszentrum Karlsruhe, Institut für Angewandte Informatik, p. 13, right
Front Design, www.materia.se , p. 55

--- G ---

Gareth Gardener/El Ultimo Grito, p. 183
Mieke Gerritzen and Karl Grandin, p. 115
George Gessert, p. 18
Martí Giuxé, p. 58, left
Bob Goedewaagen/Studio Jurgen Bey, p. 16, right
Bob Goedewaagen/JongeriusLab, p. 127, left
Golden Gate National Parks Conservancy, p. 194, left
David Goodsell, p. 94–96
Sean Godsell, p. 154
Greg Lynn FORM, p. 132, right

--- H ---

Hargreaves Associates, p. 194, right
Simone Haupt and Frank Edenhofer, p. 93
Honda, www.honda-robots.com, p. 15, left
Hooker & Kitchen, p. 153

--- I ---

Hiroshi Ishiguro at Osaka University and KOKORO Co. Ltd., p. 15, right
Institut für Mechanik und Mechatronik, Abteilung für Intelligente Handhabungs- und Robotertechnik, Technische Universität Wien, p. 14, left
Instituto Santiago Ramón y Cajal, Madrid, courtesy heirs Jorge Ramón y Cajal, p. 29
Interbreeding Field, p. 185
Iwamoto Scott mit Proces2, p. 155

--- K ---
Karlssonwilker (Jan Wilker & Hjalti Karlsson), p. 10
Angelo Kaunat, p. 133, right
Tobie Kerridge, Royal College of Art, p. 17
Barbara Klemm, p. 206, right
Kol/Mac Studio, p. 129
Krets Research Group, p. 145
K-Team, p. 14, right

--- L ---
Peter Liedtke, p. 207
Günter Lintl, p. 220, below

--- M ---
Paul McCarthy, Collection of Frac, Montpellier, France, courtesy the artist, Hauser & Wirth, Zurich, London, p. 84
Ena Macena/Nadine Meisel, p. 58, right
Map Office, p. 164–165
Matthieu Mance, Tokyo, Japan, p. 127, right
Marvel Comics, p. 81
Mass Studies, p. 147, 148–149
Thomas Mayer, p. 168, 214, 220, above, 221, 223
Anton Meinholz, p. 205
Peter Menzel, Napa, CA, USA, from the book *Robo Sapiens*, MIT Press p. 128
Catherine Mosbach/Mosbach Paysagiste, p. 192–193
Jeroen Musch, p. 189

--- N ---
Niels Bohr Archive, Kopenhagen, p. 28
Nox/Lars Spuybrock, p. 146

--- O ---
Office for Metropolitan Architecture, p. 222, 223, left
Otto Bock, p. 13, left

--- P ---
Adrian Pingstone, p. 63
Ezio Prandini, Campeggi, p. 54
PTW Architects, CCDI, ARUP and PTW Architects Chris Bosse, p. 12

--- R ---
R&Sie(n), p. 150
©Albert Renger-Patzsch Archiv, Ann u. Jürgen Wilde, Köln 2006, ProLitteris, Zurich, p. 202, 203
Chris Roberts, p. 190

--- S ---
Lothar Schnepf, p. 83
Joachim Schumacher, p. 168
Sotheby's, p. 62

--- T ---
Timorous Beasties, p. 178

--- U ---
Universitätsbibliothek Salzburg, http://www.ubs.sbg.ac.at, p. 78
Universität Stuttgart, TGU Napshell, www.napshell.com , p. 66

--- W ---
Warner Brothers, p. 80
Jack van Wijk, p. 116–117

PUBLICATION INFORMATION

Roger M. Buergel, "Der Schönheit eine Chance" was first published in *Die Zeit*, April 14, 2005, No. 16, p. 53. Permission for reprint was kindly given by the author and *Die Zeit*.

Vilém Flusser, "The House of the Net" was first published in English as: "With as many wholes as a Swiss cheese" in: Vilém Flusser, *The Shape of Things*, Reaction Books, London 1999. Reprint courtesy by Edith Flusser.

Bill Joy, "Why the future doesn't need us" was first published in *Wired*, April 8, 2000.

Peter Sloterdijk, "Das Zeug zur Macht" is the edited version of the public lecture given by Sloterdijk as part of the International Design Forum in Ulm on September 22, 2005, under the heading "Transformation".